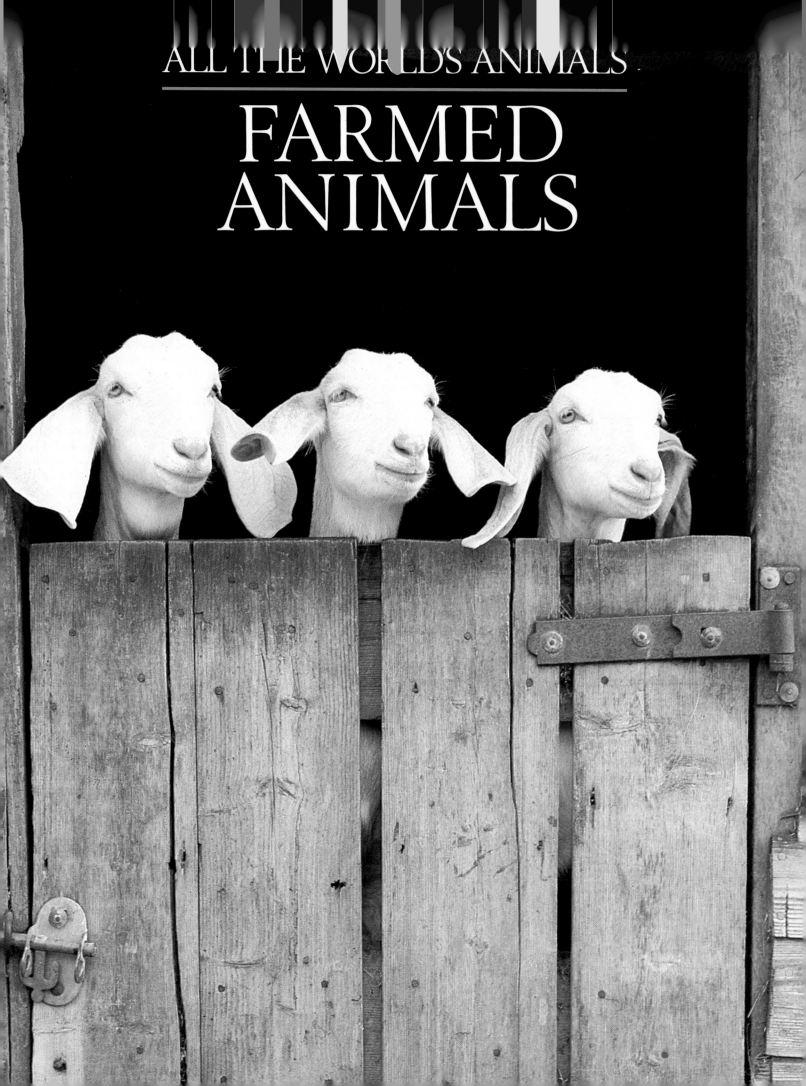

ALL THE WORLD'S ANIMALS

FARMED
ANIMALS

ALL THE WORLD'S ANIMALS

FARMED ANIMALS

TORSTAR BOOKS
New York · Toronto

CONTRIBUTORS

JLA Jack L. Albright
Purdue University
West Lafayette, Indiana
USA

MAB Miles A. Barton
University of Reading
Reading
England

BCRB Brian C. R. Bertram
Zoological Society of London
London
England

DMB Donald M. Broom
University of Reading
Reading
England

AD Alexandra Dixon
International Council for Bird
Preservation
Cambridge
England

PD Patrick Duncan
Station Biologique de la Tour
du Valat
Arles
France

SAE Sandra A. Edwards
Terrington E.H.F.
King's Lynn
England

YE Yngve Epsmark
Universitetet Trondheim
Trondheim
Norway

JFl John Fletcher
Auchtermuchty, Fife
Scotland

JBF John B. Free
Rothampsted Experimental
Station
Harpenden
England

RG Ray Gambell
International Whaling
Commission
Cambridge
England

SJGH Stephen J. G. Hall
University of Cambridge
England

EHi Eric Hillerton
University of Reading
Reading
England

AM Alan Mowlem
The Animal and Grassland
Research Institute
Shinfield, Reading
England

PLN P. Le Neindre
Institut Nationale de la
Recherche Agronomique
Theix, Beaumont
France

LJP Lynnette J. Peel
University of Reading
Reading
England

MWR Matt W. Ridley
University of Oxford
England

HHS Hans H. Sambraus
Universität München
Freising-Weihenstephan
West Germany

CJS C. Jonathan Shepherd
PH Pharmaceuticals Ltd
Stanmore, Middlesex
England

JLS John L. Skinner
University of Wisconsin
Madison, Wisconsin
USA

CET Clair E. Terrill
US Department of Agriculture
Beltsville
USA

TTT Timothy T. Treacher
The Animal and Grassland
Research Institute
Hurley, Maidenhead
England

GFW George F. Warner
University of Reading
Reading
England

JMW J. M. Wilkinson
Marlow, Bucks
England

JWi Julian Wiseman
University of Nottingham
Sutton Bonington
England

DY David Yerex
Masterton
New Zealand

ALL THE WORLD'S ANIMALS
FARMED ANIMALS

TORSTAR BOOKS
300 E. 42nd Street,
New York, NY 10017

Project Editor: Graham Bateman
Editor: Stuart McCready
Art Editors: Chris Munday, Jerry Burman
Art Assistants: Wayne Ford, Carol Wells
Picture Research: Alison Renney
Production: Clive Sparling
Design: Nikki Overy
Index: Stuart McCready

Originally planned and produced by:
Equinox (Oxford) Ltd
Littlegate House
St Ebbe's Street
Oxford OX1 1SQ
England

Editor:
Dr Donald M. Broom
The University of Reading
Reading
England

Advisory editors:
Professor Jack L. Albright
Purdue University
USA

Professor Dr Hans H. Sambraus
Lehrstuhl für Tierzucht der
Technischen Universität
München
West Germany

Artwork panels:
Priscilla Barrett
Mick Loates

On the cover: Highland bull
Page 1: Goat kids
Pages 2–3: Cattle
Pages 4–5: White leghorn
Pages 6–7: Cattle round-up
Pages 8–9: Turkeys

10 9 8 7 6 5 4 3 2 1

Printed in Belgium

Library of Congress Cataloging in Publication Data

Farmed animals.

  (All the world's animals)
  Bibliography: p.
  Includes index.
  1. Animal Culture. 2. Livestock. 3. Domestic animals. I.
Series.
SF65.2.F37 1986   636   86–16165
ISBN 0–920269–86–9

ISBN 0-920269-72-9 (Series: All the World's Animals)
ISBN 0-920269-86-9 (Farmed Animals)

In conjunction with *All the World's Animals*
Torstar Books offers a 12-inch raised
relief world globe.
  For more information write to:
Torstar Books
300 E. 42nd Street
New York, NY 10017

CONTENTS

FOREWORD

The commonest large mammals in the world are domestic cattle, sheep, pigs goats and buffalo, and the commonest bird in the world is the domestic chicken. These, along with bees, silk moths, trout, oysters and many other farmed animals are the subject of this fascinating and revealing book.

The development of civilization has depended on the farmed animals and almost all societies depend on their use today. These creatures supply man with food (meat, milk, eggs, fish) and with basic protection against the elements (furs, hides, wool for clothing). They are used as working animals and beasts of burden. Some, such as certain ducks, geese and other fowl, are bred purely for ornamental reasons, while others provide some of the luxuries of life, from quails' eggs to cashmere.

Animals have been farmed from very early in man's history. The bones of obviously "domesticated" sheep and goats, which can be dated to more than 10,000 years ago, have been found in the remains of settlements in southwest Asia. More recently, the Old Testament, Exodus 26:7–13 (written *c.* 1000BC) includes a reference to the weaving of fine goat hair into curtains.

Humans often display an affectionate attitude toward animals kept for useful purposes. The docile water buffalo, popularly described as "the living tractor of the East" may be "pensioned off" after 20 years' work to live in retirement as one of the family. Apart from those who work with them or make special studies of them, most people are inclined to take farmed animals for granted. Yet recent research shows that they often display a complexity and sophistication of behavior that commands respect. Members of a herd of 40 cows, for example, will recognize and respond to each other as individuals, and the "contented" but not particularly "intelligent" cow is surprisingly adept at picking up the complex tasks involved in modern automated feeding methods.

The fascination of farmed animals as subjects for closer study becomes abundantly clear from the enthusiasm conveyed by the authors of this book. The informative, in-depth treatment is complemented by the diversity of the illustrations. This combination can only enhance and enrich the reader's perception and understanding of farmed animals.

How this book is organized

Farmed Animals includes articles by specialist authors from six different countries, on the animal species farmed by man throughout the world, and on the methods used in the production of important commodities. Pains have been taken to gather up-to-date statistical data giving a world picture of products and producers, and to present these in readily accessible charts and diagrams.

The rearing of animals for the conversion of animal food to meat, and so on, and some of the procedures involved in the production of these commodities, are discussed in detail in the first article. The theme of the value to man of farmed animals is pursued in each species article in a boxed summary of the world importance of that species. As well as considering what man gains from farmed animals, our obligations to them are also discussed. The necessity to take into account the welfare of each individual animal is emphasized in that part of the article on ethics and animal husbandry (see pages 22–23). Public attitudes toward preserving the variety of animal life are reflected in discussions of the conservation of rare breeds, the ways in which the wild populations, such as fishes and whales, should be treated (see pages 126–139) and in a forward-looking article on farming endangered species (see pages 150–151). Man's ingenuity in using animals to collect food and other materials is illustrated by many intriguing examples relating to hunters and gatherers (pages 146–149).

Space is devoted to the various species in proportion to their importance to man and the diversity of their breeds. Each species is introduced by an opening panel giving details of classification, distribution, physiology, diet and reproduction. In the main articles are details of each animal's biology and a survey of the full breadth of its place in human life. Authors explain what is known of the origin of the animal's domestic populations and the historical development of its role. Where possible, insights are given into the natural pre-adaptations of each for a domestic niche. Separate fact box summaries following the main articles on major species give the history and characteristics of many of the breeds that have resulted from the domestication process. The great diversity of breeds adapted to physical conditions and husbandry practices in the various regions of the world is emphasized in particular.

The recent increases in activities of rare breed societies in many countries, and the proliferation of farm parks, are evidence of the continuing interest of the scientific community and the general public in old and local breeds, as well as those that are the most successful in modern farming.

THE PRODUCTS

Farmed animals make a smaller contribution than plants to a hungry world. Less than a quarter by weight of world food production comes from animal products, and food energy present in plants is lost during conversion to animal products. Animals do, however, provide a much higher proportion of the total human protein intake (mainly from meat) and they allow man to utilize many plants which are not themselves suitable for human consumption. Vast areas of grassland are harvested by farm animals and production of plants in freshwater and in the seas is converted by fish into valuable protein which man consumes. Man also needs animal products for clothing, shoes, bedding and other means of protecting himself from his environment. The total world production of hide, wool etc used for such purposes is about 10 million tons per annum.

Some 600 million buffalo, cattle, asses and mules etc are used as sources of power, compared to 21 million tractors. Draft animals produce much more energy input to agriculture than do tractors—estimated at nine times as much in 1980.

World food production
- Eggs 30
- Fish etc 75
- Animal Products
- Meat 144
- Other Cereals 340
- Wheat 460
- Dairy Products (mainly milk) 483
- Rice 410
- Potatoes 260
- Other Roots 300
- Corn 450
- Fruit 280
- Other Vegetables 350

◀ **World food production** (million tons per annum). Although animal products include water in milk, they provide more protein than crops. Not all crops are for human consumption.

▼ **Cattle market at Dodoma, Tanzania.** These humped zebu cattle, originally from India and Pakistan, are resistant to tropical temperatures and need less water than European cattle. They are widespread in Africa.

MEAT, MILK, WOOL...

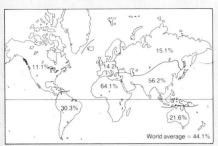

World average = 44.1%

Total world production: 734 million tons per annum*
Figures are in million tons

Meat		Eggs	
World	144.7	World	29.8
Africa	7	Africa	1.0
N and C America	30	N and C America	5.4
S America	11.9	S America	1.8
Asia	37.4	Asia	10
Europe	39.2	Europe	7.3
Australasia	3.9	Australasia	0.3
USSR	15.3	USSR	4
Milk		**Fish**	
World	483	World	74.7
Africa	13.1	Africa	4.0
N and C America	80.4	N and C America	7.2
S America	24.3	S America	8.6
Asia	75.4	Asia	31.5
Europe	187.3	Europe	12.5
Australasia	12	Australasia	0.4
USSR	90.5	USSR	9.5
Wool			
World	1.72		
Africa	0.1		
N and C America	0.04		
S America	0.17		
Asia	0.27		
Europe	0.16		
Australasia	0.71		
USSR	0.27		

*Based on 1982 FAO statistics, except 1981 Fisheries

FARMED animals are kept primarily for their meat; although milk for human consumption is an important product, historically it has been secondary to the rearing of animals for slaughter. In areas where dairy herds are kept for milk production, beef and veal are by-products from male calves and from female calves which are surplus to the requirement for herd replacements. In addition, cows which are slaughtered at the end of their productive life are a significant source of meat.

Meat

The two main methods of keeping meat animals are confinement in houses or grazing on pasture land. Pigs and poultry are normally confined, often with a degree of control over the natural environment. In particular, the houses are ventilated and the temperature is kept relatively constant.

Grassland comprises 20 percent of the land area of the world, and it is here that the majority of the beef, mutton, lamb, goat meat, buffalo meat and horse meat is produced. Breeding females graze in herds with a few males and rear young until these are either weaned or slaughtered. The yield and quality of grassland is relatively low compared to that used for arable cropping and as a result the rate of growth of the young can be limited by both a low yield of milk from its mother and a low availability of nutrients from the pasture itself.

Grassland supplies most of the nutrients of the cattle, sheep and goats kept for meat. Cereal grains are the major sources of feed for pigs and poultry. In areas of the world where grain production exceeds the local demand for human consumption, for example the United States, beef cattle are given mixtures of grain and forages such as corn silage and alfalfa (lucerne). The animals are kept in large feedlots which are mechanized to the extent that one man can feed up to 10,000 cattle each day.

Cereals are of high energy value and promote rapid rates of animal growth. Thus the life span of animals given diets based on cereal grain tends to be relatively short, ranging from 2 months in the case of broiler chickens to 11 months for beef calves reared on barley as the main feed from weaning at 3 months of age to slaughter. By contrast, cattle reared on pasture may be two years old or more before they achieve a heavy enough weight to be slaughtered.

Meat animals add value to feeds, but when they receive grains and protein-rich feeds such as soybeans they compete with man for nutrients. The ruminants (cattle, sheep, goats and buffalos) play a vital role in producing human food from uncultivatable land which would otherwise be unproductive. Animals also consume by-products such as straw, husks and wastes from the food industry.

Breeding pigs may have as many as 20 offspring in two litters each year. Like chickens hatched from eggs laid by a small number of selected breeding females, pigs carry a relatively low "overhead" cost in terms of the requirements of the mother for land and for feed. In contrast, cattle, buffalos and horses carry a much larger overhead cost. The female must be kept for an entire year, during which time she may rear only one offspring. Sheep and goats are intermediate in

their reproductive performance, though in the more arid areas of the world they are often limited by inadequate nutrition.

The farmer produces a live animal which, when judged fit for slaughter or for culling from the herd or flock, is processed for human consumption. At the slaughterhouse it is stunned, bled, beheaded, eviscerated, skinned and sawn in half. Poultry and lambs are normally sold whole after the head, feet, viscera and feathers or skin have been removed. Before leaving the slaughterhouse the meat is inspected to ensure that it is fit for human consumption. The liver and lungs of each animal are examined for evidence of diseases which might contaminate the meat.

In temperate climates the carcass is normally hung at low temperature, before being cut up. In hot climates the meat is normally butchered and consumed immediately after slaughter to avoid decomposition.

Beef and horse meat are normally sold without bones. Pig meat, lamb, goat meat and poultry meat are sold and cooked on the bone. The proportion of the carcass which is edible varies with the amount of meat in relation to the amount of bone and trimmed fat, but would typically be about 70 percent. Meat is sold in small pieces which correspond to the musculature or to the limb structure of the animal, depending on different butchery customs. The object is to maximize the value of the carcass and at the

▼ **Confined to the poorest grazing,** goats make a relatively small contribution to world meat totals, but they are efficient at converting feed to meat. Many, like these goats on the banks of the River Gambia in West Africa, make valuable use of land which cannot sustain crops or other grazers. Over 300 million goats are reared for meat in Africa and Asia, with skins a major secondary product. In cooler climates wool and dairy products are also important. Goat meat is a by-product of a small specialty dairy industry in western countries.

same time provide the consumer with meat in suitably sized "packages" for cooking. A typical piece of beef for roasting, for example, would weigh 3.31lb (1.5kg); an individual portion of pork or beef for grilling would weigh 7oz (200g).

Meat is primarily a source of protein in the diet. A portion of cooked meat of 3.5oz (100g) supplies between 5 and 10 percent of the recommended daily intake of energy of an adult male person. The same portion will also supply between 30 and 45 percent of the recommended daily intake of protein. Beef and lamb are significant sources of iron, but the contribution made by meat to the requirements for other minerals and for vitamins is low.

Meat supplies its energy mainly from fat which is distributed within and between lean muscle tissue. Pig meat, especially pork sausage, is the major contributor of animal fat to the diet in developed countries, not only because it has a relatively high concentration of fat but also because pig meat is a cheap and popular food.

Depending on grade, type, diet, and condition of the carcass, lamb, like pork, may have 30 percent fat. Beef contains less fat (24 percent) while chicken (14 percent), rabbit (8 percent) and turkey (7 percent) are considerably leaner meats.

Worldwide, the contribution of meat to dietary energy supply is low, but it supplies almost one-fifth of the protein intake of the world's population. Daily intake varies from an excess of protein in Australia and New Zealand to as little as 2 percent and 7 percent of the recommended intakes of energy and protein respectively in Asia.

There are good reasons for encouraging an increase in meat consumption in areas of the world where protein is in very short supply and where the population is undernourished. On the other hand, there is a strong case for a reduction in the consumption of animal fat in countries where meat is eaten in large daily amounts. This is because there is evidence of a link between the amount of animal fat in the diet and heart diseases. Reductions in intake of fat can be achieved by eating less processed meat such as sausages, and by choosing meat from leaner animals and from birds.

JMW

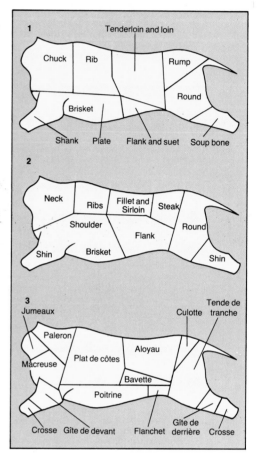

▼ **Beef cuts.** Cuts of meat vary from country to country. Shown here are the cuts of beef used in (1) United States, (2) United Kingdom, (3) France.

World Production of Meat

The pig is the most important meat producing species in the world. Almost 800 million are slaughtered annually to produce 55 million tons of meat. Pig meat (see p93), together with meat from cattle and poultry, accounts for over 90 percent of world production. Sheep meat (see p68) and goat meat (see pp86–87) are of less importance; the animals tend to occupy the less productive grasslands of the world and in consequence their yields of meat tend to remain low. Buffalo meat is produced mainly in China and, to a lesser extent, in Egypt. Much of the world's horse meat is produced in China and Italy. Argentina, the United States and Brazil are the major exporters of horse meat.

The most important meat producing areas of the world are Europe, Asia and North and Central America. Both Europe and North America have levels of production of around 132lb (60kg) of edible meat per person per annum. In Asia, with its vast population of over two billion people, the output of edible meat is only 24.2lb (11kg) per person each year. South America and the USSR have similar levels of production. Meat is exported from South America, but the USSR is a net importer. The production of meat in Africa is low, both in relation to its total area of land and per head of population. In Oceania, production is also relatively low, but very high in relation to its human population. Consequently, both Australia and New Zealand rely on the export of meat as a significant source of income.

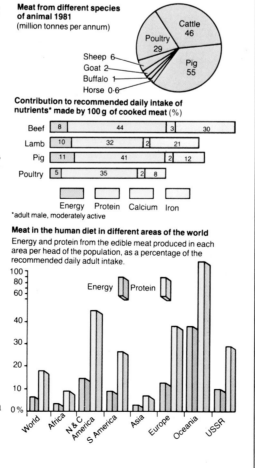

Meat from different species of animal 1981
(million tonnes per annum)

Cattle 46
Poultry 29
Pig 55
Sheep 6
Goat 2
Buffalo 1
Horse 0·6

Contribution to recommended daily intake of nutrients* made by 100 g of cooked meat (%)

	Energy	Protein	Calcium	Iron
Beef	8	44	3	30
Lamb	10	32	2	21
Pig	11	41	2	12
Poultry	5	35	2	8

*adult male, moderately active

Meat in the human diet in different areas of the world
Energy and protein from the edible meat produced in each area per head of the population, as a percentage of the recommended daily adult intake.

Energy   Protein

World, Africa, N & C America, S America, Asia, Europe, Oceania, USSR

▶ **As many as 10,000 beef cattle at once** can be reared by one man in some automated North American feed lots. The animals are nourished on rich rations of grain, alfalfa and corn as well as hay such as that provided for these calves, mainly Herefords, in Brawley, California. Feed-lot animals that never see a pasture may be ready for market at 11 months, compared to two years or more for the grazing cattle which produce most of the world's beef.

Milk

Milk has been studied in overwhelming detail in seven species: the cow, water buffalo, goat, sheep, horse, pig and man. Miscellaneous and scattered data are available on about 150 other species. But it is one animal, the dairy cow, that produces the great bulk of milk consumed by humans in various liquid and solid forms.

The major portion of cow's milk—87.3 percent—is water. It is a fluid carrier for the other basic constituents: milkfat—also called butterfat—(3.7 percent) and nonfat solids (8.9 percent). The nonfat solids include 3.4 percent protein, 4.8 percent lactose (milk sugar) and 0.7 percent minerals. These percentages vary with the breed and stage of lactation of the cow.

Butterfat is the most variable constituent of milk. It contains fatty acids essential to the human diet and adds flavor and body to milk. Prices of milk to the farmer are paid largely on volume weight and adjusted on percentage butterfat.

Research studies have concluded that approximately 60 percent of the variation in milk composition among cows is due to inherited traits. Animals with a high ability to produce butterfat tend also to be good protein producers. The environment, in

particular feed, accounts for the remaining 40 percent variation.

A high grain ration (less than one-third roughage) depresses butterfat and slightly elevates protein in milk. Insufficient energy slightly reduces nonfat solids and their protein content and greatly reduces milk yield.

Hot weather tends to depress both butterfat and nonfat solids. Both of these are higher in early and late lactation. After ten months in production, a cow's yield drops and it is more efficient to let her go dry and have her calve again. During the dry period the cow's body and mammary system undergo rest and regeneration. The longest nonstop lactation period in history without the interruption of having a calf belongs to a pedigree cow called "Old Jersey." She produced for at least 15 years without a break.

Mastitis (an infection of the udder) and other diseases can decrease the total of nonfat solids and make the milk susceptible to off-flavors. Because mastitis changes the composition of milk and lowers overall production, products manufactured from mastitic milk will be lower in quality and yield.

The flavor of fresh fluid milk is difficult to describe because it is a bland product. It can leave either a slightly sweetish taste because

▲ ▼ **"The foster mother of the human race"** is traditionally milked where she happens to be, as ABOVE in a village street by a boy of one of the Dinka tribe of Southern Sudan. BELOW Milking machines are used in fixed milking parlors, such as this "abreast" parlor. (See also p57.) Cows and buffalos are still milked by hand in the streets of Calcutta and Cairo, and up until recent times cows were sometimes milked in fields and city streets in Europe and North America.

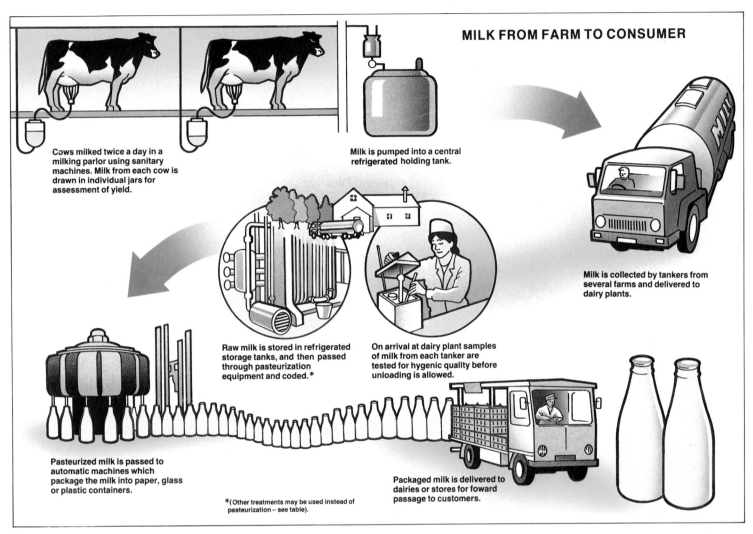

MILK FROM FARM TO CONSUMER

Cows milked twice a day in a milking parlor using sanitary machines. Milk from each cow is drawn in individual jars for assessment of yield.

Milk is pumped into a central refrigerated holding tank.

Milk is collected by tankers from several farms and delivered to dairy plants.

Raw milk is stored in refrigerated storage tanks, and then passed through pasteurization equipment and coded.*

On arrival at dairy plant samples of milk from each tanker are tested for hygenic quality before unloading is allowed.

Pasteurized milk is passed to automatic machines which package the milk into paper, glass or plastic containers.

Packaged milk is delivered to dairies or stores for foward passage to customers.

*(Other treatments may be used instead of pasteurization – see table).

of its lactose content, or a slight suggestion of salt due to the presence of chloride salts. Fresh milk should impart a pleasant smooth sensation to the mouth and there should be no evidence of astringency. When the intensity of the normal flavor changes or other flavors appear, the consumer objects. Feed flavor can enter the milk from feeding silage or green feed just before milking, and cowy or barny flavors can be caused by a poorly ventilated environment. Other flavor defects can result from bacterial activity. All these are under the direct control of the milk producer.

Animal feed to milk. In making milk, cells of the mammary gland (the secreting tissue in the udder) take water, glucose, amino acids and other compounds from the blood. It has been estimated that blood makes a complete cycle from the udder to the heart and back in less than a minute (52 seconds). About 1,100lb (500kg) of blood are needed 2.2lb of milk secreted in a cow or goat. A cow producing 110lb (50kg) of milk per day would need to pump about 55,000lb (25,000kg) of blood through her udder daily. The optimal availability of raw mat-

erials for milk in this flow requires sound feeding of the dairy animal.

When modern dairy farmers (see pp56–57) want to know correct amounts and combinations of feed, they can be calculated by a computer programmed specifically for the nutritional needs of the herd. Dairy scientists have learned that a growing replacement heifer (female yet to produce milk) destined to be a cow might require a ration that has only 12 percent protein, while a high producing cow might need more than 16 percent protein. Early cut grasses and prebloom alfalfa (lucerne) hay have over 20 percent protein. Corn that is cut green, chopped and stored as dry corn silage in silos contains about 8 percent protein.

The better the diet, the more likely it is that a cow will eat to meet her nutrient requirements. Each day a productive cow offered a good diet can eat dry matter at the rate of 3 to 4 percent of her body weight (an average black and white Holstein-Friesian weighs about 1,500lb–680kg). A similar cow on poor feed might take in only 1.5 percent.

Types of Milk

Untreated
Raw milk not heat treated.

Pasteurized
Milk heated to not less than 162°F (72°C) for at least 15 seconds and then rapidly cooled. This kills bacteria and improves keeping quality, without reducing nutritional value.

Ultra Heat Treated (UHT)
Milk heated to at least 270°F (132°C) for one second and packed rapidly under sterile conditions. Keeps for several months in unopened container, even without refrigeration, with minimum effect on flavor.

Sterilized
Milk heated in filled bottles to 219–235°F (104–113°C) and left to cool naturally. Keeps for two to three months in unopened container, without refrigeration, but there is often a cooked flavor and more creamy appearance.

Homogenized
Milk sprayed through tiny openings to create smaller fat globules, thus stopping formation of cream layer. Creamier tasting and more easily digested than ordinary milk.

Milk Producers of the World

Cows are the primary milk producers serving mankind. They produce about 91 percent of the total milk supply, followed by Water buffalos with 6 percent and sheep and goats with 3 percent. In 1983 the annual world production of milk for human consumption was estimated at 438 million tons from cows, 28 million tons from Water buffalos, 8.1 million tons from sheep and 7.7 million tons from goats.

The total of sheep's milk (see p69) has been increasing moderately, especially in Europe. Cheese made from sheep's milk, including the original Roquefort cheese produced in southern France, is very popular in many countries. In recent years, goat's milk (see p83–86) has made modest gains, and in many underdeveloped areas goats are still the main livestock serving human needs. Of the estimated 400 or more million goats, about three-quarters, mainly meat animals, are found in African and Asian nations.

Nomads still use camel's milk (though Saudi Arabia is now developing a dairy industry based on the cow). Reindeer milk and cheese are still food items in parts of Siberia. Humped zebu cattle (see pp44–47) are the main milk producers in India and Central Asia. Water buffalo milk (see p60) is important in India and Egypt, yak's milk (see p62) in Tibet. The llama is used in South America and the mare in Asia.

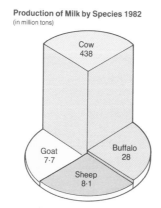

Production of Milk by Species 1982 (in million tons)

Cow 438 · Goat 7.7 · Buffalo 28 · Sheep 8·1

Compositon of Fresh milk

Major Constituents (% of fresh milk)	Ewe	Cow	Goat
Total solids	18.4	12.1	13.2
Solids-not-fat	10.9	8.6	8.7
Fat	7.5	3.5	4.5
Protein	5.6	3.25	3.3
Lactose (milk sugar)	4.4	4.6	4.4
Minerals (ash)	0.87	0.75	0.80
Calcium	0.19	0.12	0.14
Phosphorus	0.15	0.10	0.12
Chlorine	0.14	0.11	0.15
Energy concentration (MJ/kg of fresh milk)	4.4	2.6	3.0

Average composition of commonly used dairy products

Fat · Protein · Lactose · Minerals (ash) · Water · Calories (lb)

Product	Calories (lb)
Whole milk	310
Skim milk	162
Buttermilk	167
Cheese (cottage)	454
Cheese (Swiss)	1832
Cheese (Cheddar)	1916
Cheese (cream)	1909
Evaporated milk	629
Condensed milk	1484
Dried whole milk	2247
Dried skim milk	1632
Cream	1909
Butter	3325

During 1975, her world-record year for milk production (55,661lb—25,247kg), Beecher Arlinda Ellen, an American Holstein, averaged 5.5 percent dry matter intake as a percentage of her body weight. Her highest single day's yield was 195.5lb (88.9kg). While the human record may be close to 13lb (5.9kg), the human average is closer to 2.2lb (1kg).

Milk in nutrition. Milk is considered to be nature's most nearly perfect food. Milk and dairy foods are significant sources of several nutrients, particularly calcium, riboflavin, phosphorus, protein, magnesium, vitamin B12, niacin, vitamin B6 and, when fortified, vitamins A and D. Rickets, which was commonly seen as late as the 1930s, almost vanished from the developed world with the legal requirements to fortify milk with vitamin D.

Milk proteins provide all the essential amino acids to build and replace tissues, muscle, blood, skin, hair and hormones. For its content of calcium and other nutrients important in bone formation, milk and dairy foods may be of particular value in prevention, mitigation and reversal of age-related diseases affecting bone density.

Pastoral peoples of India and Africa who subsist on diets containing a large proportion of milk, curdled milk, cheese and fermented milks are healthier, better-developed and longer-lived than those living on cereals.

The high nutrient content of dairy foods when compared to the calories they contain makes these foods an economical source of nutrients for the calorie- and weight-conscious consumer. The consumer has a wide range of dairy foods from which to choose: whole milk, low-fat milk, skim or nonfat milk, flavored milks, evaporated and condensed milks, dry milk; several cream products; cultured milk and culture-containing dairy products like buttermilk, sour cream, yogurt; speciality milks such as low-sodium milk required for reasons of health and lactose-hydrolyzed milk for lactose-intolerant consumers (see below); more than 500 different cheeses; butter; ice cream and related frozen desserts.

Yet the place of milk in a nutritionally sound diet is under debate. Milk is naturally low in iron and vitamin C. Dairy foods have also been singled out as undesirable because of their cholesterol and saturated fatty acid

▲ Too precious to be used as manure.
A valuable by-product of draft and dairy cattle in India is their dung. Shaped into cakes and dried, it supplies cooking fuel, leaving ashes for fertilizer. Some villages now collect cowpats in central fermenters which produce piped methane gas. The gas can be burnt with greater fuel efficiency than dung cakes and a richer residue which is easier to collect than cooking ashes can be returned to the fields.

◄ From a barely visible undercoat beneath the coarse outer hair of the first domestic sheep, man has selectively bred over thousands of years for the modern fleece, here being shorn in New Romney, Kent, England. Wool breeds do not molt; instead coats are retained as tidy parcels to be shorn off. The coats of some breeds now consist purely of wool. White is favored because it can be dyed easily. Although wool has been historically the most important product from sheep, single-purpose meat breeds predominate today.

content. The data, however, are inconclusive, and recently it has been shown that milk as well as yogurt exhibit a cholesterol-lowering effect; the exact mechanism is undefined.

Controversy also exists over lactose intolerance, which has been pointed to as a reason against supplemental milk feeding programs aimed at children. Some people are unable to metabolize milk sugar (lactose) because they have a low level of the enzyme lactase. As children grow older and milk consumption decreases, the amount of lactase in their gastric systems gradually decreases to some extent. The result can be lactose intolerance. It is high (60–100 percent) in most non-white populations and low (0–35 percent) among relatively few of the world's white populations. It has been suggested that only among descendants of African and northern European populations that have raised dairy animals since antiquity do people retain beyond childhood the ability to produce lactase.

When given a 1.7oz (50g) test dose of lactose, intolerant individuals may show an inadequate rise in blood glucose along with gastric symptoms which may include gas, bloating, cramps, flatulence and diarrhea. That amount of lactose would be found in 1.075qt (1 liter) of milk.

Research has shown, however, that even individuals who develop symptoms with large test doses can consume small recommended amounts of milk at intervals in the day. There are only 0.42oz of lactose (12g) in a 8oz (225g) glass of milk. Fermented milk products such as yogurt and lactose-hydrolyzed milk have also been suggested as sources of calcium for lactose-intolerant individuals. Lactose intolerance need not imply intolerance to milk.

Lactose intolerance is sometimes confused with other ailments which are very different medically but whose names sound similar. Milk allergy affects less than 5 percent of infants and children under two years of age. These children exhibit an allergic response to the protein in milk just as some people are allergic to pollen grains. Such milk allergies are usually outgrown by two years of age. For those who are truly intolerant to cow's milk, suitable alternatives include most cheeses and fermented dairy products and goat's milk. JLA

Wool, Hide, Fur, and Feathers

The skins of farmed animals and the products which they grow on them—wool, fur and feathers—have often been economically just as important as food: cattle have sometimes been reared primarily for their hides, and during most of the history of domesticated sheep, meat was a by-product of wool production.

Today skins may be the primary product of crocodile farms (see p150), furs of mink farms and feathers of ostrich farms, but all of the major farmed animals are now food producers first and foremost. Even wool is now mainly a by-product: synthetic substitutes have cut deeply into its market, and sheep farmers are committed to supplying a higher consumer demand for meat than the world has ever known before. However, as food production increases to meet the needs of an expanding world population, so inevitably does the supply of by-products, and their economic importance continues to be considerable.

Wool. Camel hair blankets are a common item in bazaars and markets around the fringes of the great deserts plied by the dromedary. Alpacas (see p64), Angora rabbits, some producing up to 14oz (400g) in a year, and several breeds of goat (see pp88–89) are reared for their fine wool. But it is sheep that are the overwhelmingly most important producers of animal fibers.

Sheep (see pp66–79) produce three types of fiber: wool, kemp and hair. Wool fibers range in diameter from 0.0006–0.0016in (15–40 micrometers). Coarse wool fibers may have a fine hollow core, or medulla, along part of their length but fine wools have no medulla. Kemp, with a diameter of more than 0.004in (100 micrometers), has a wide medulla with a latticed appearance running the whole length of the fiber. Hair fibers are intermediate between wool and kemp with a discontinuous medulla. The coarser fibers are produced by large primary follicles, which have an associated sweat gland and erector muscle. Wool grows from smaller secondary follicles grouped around the primary ones.

The type of wool produced depends on the breed of sheep but both the animals and the wool must be adapted to climate and feed conditions. This is not a problem with native or well-established breeds where selection to produce the desired type of wool has taken place over time as the breed became adapted to the climate and feed.

The Merino breeds originating in Spain are superior for the production of fine wool, and have spread all over the world. Longwool breeds (Romney, Texel, Leicester, Lincoln), Down breeds (Hampshire, Southdown, Suffolk) and crossbred sheep (Corriedale, Polworth) are kept in good pastures, primarily for meat production. The wool is very useful for general purposes such as blankets, coats and a variety of clothing and textile materials. The longwool breeds are kept on the lushest pastures while the crossbred sheep may be kept on sparser vegetation or on semiarid range.

Long wool is also used for carpets and mattresses as is wool from fat-rump breeds (Blackhead Persian), fat-tail breeds (Awassi) and carpet wool breeds (Scottish Blackface, Churro). Coarse wool is well suited to hand processing as are medium and colored wools.

Wool may prevail over meat production in warm, dry climates where sheep are grazed the year round and often under sparse feed conditions. In mountainous areas the sheep may graze at different elevations in different seasons—in the lowlands and valleys, sometimes with irrigated fields, in winter; in foothills in spring and fall, and in high mountains in midsummer. Lambing and shearing occur in late winter, spring or early summer. Lush vegetation is better suited to the higher reproductive rates and rapid growth of efficient meat production. Sheep kept primarily for wool are rarely housed indoors except for temporary protection during lambing and shearing.

Hides. Cattle hides provide the main source of leather for a wide variety of products—

▲ **Sheep grazing** on rape which is a valued forage crop used for late fall and early spring pasturing. The seed is often sown after a cereal crop.

▶ **Tanning leather.** ABOVE a chart of the stages in the industrial production of leather. BELOW Tanners' district in Marrakesh, Morocco. The vats hold different liquids for different stages of the process, including water to soak fresh hides, urine to remove hair and wool, and tannin to change skin proteins.

World Production of Wool

Wool production is increasing in the world but very slowly. It totaled 1.7 million tons in 1982. Australasia produces over a third, followed by Asia, South America, Europe and Africa. Production trends are still upward in all except the Americas. Ten leading countries produce 80 percent of the world's wool. Australia is by far the leader. The USSR is second but now showing signs of decline, and New Zealand, third, has leveled off.

Wool produced per sheep is declining in the world at the rate of 2 percent per annum. Emphasis is shifting from wool to meat because of rapidly increasing human populations and the availability of wool substitutes. The average per capita production of wool in the world was 13oz in 1982, down from 16oz in 1961–65.

As the proportion of farm income from sheep meat increases relative to wool, wool production receives less attention. In time wool may become everywhere a by-product of meat as it already has in the western world. However, wool production will probably increase with increasing sheep numbers, especially in temperate and cold climates where wool is essential for protection of animals reared in large numbers for meat.

Trends in Wool Production

	Total Clean Wool 1961–65	1982 (1,000 tons)	Number of Sheep 1961–65	1982 (millions)	Clean Wool per Sheep 1961–65	1982 (lb)
Europe	147	162	134	143	2.4	2.4
N & C America	63	35	37	23	3.7	3.3
Australasia	648	705	211	212	6.8	7.2
Asia	144	270	238	343	1.3	1.7
Africa	96	101	140	186	1.5	1.1
S America	188	173	121	108	3.5	3.5
Australia	443	436	161	138	6.1	7.0
USSR	217	270	134	142	3.5	4.1
New Zealand	205	269	51	74	8.8	7.9
China	37	120	65	109	1.3	2.4
Argentina	99	80	48	30	4.6	5.9
South Africa	68	53	39	32	3.7	3.7
Uruguay	49	46	22	23	4.8	4.4
UK	39	40	30	33	2.8	2.6
Turkey	24	34	33	50	1.5	1.5
Pakistan	11	26	11	31	2.2	1.7
World	**1,503**	**1,716**	**1,016**	**1,158**	**3.3**	**3.3**

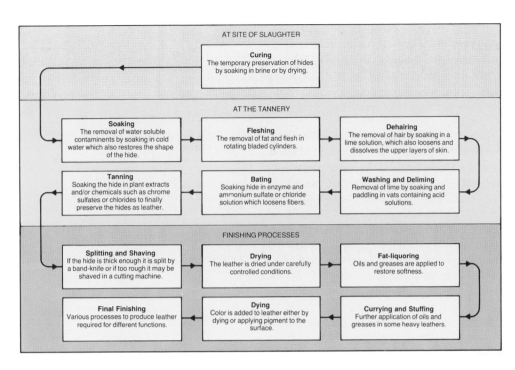

AT SITE OF SLAUGHTER		
	Curing The temporary preservation of hides by soaking in brine or by drying.	

AT THE TANNERY		
Soaking The removal of water soluble contaminents by soaking in cold water which also restores the shape of the hide.	**Fleshing** The removal of fat and flesh in rotating bladed cylinders.	**Dehairing** The removal of hair by soaking in a lime solution, which also loosens and dissolves the upper layers of skin.
Tanning Soaking the hide in plant extracts and/or chemicals such as chrome sulfates or chlorides to finally preserve the hides as leather.	**Bating** Soaking hide in enzyme and ammonium sulfate or chloride solution which loosens fibers.	**Washing and Deliming** Removal of lime by soaking and paddling in vats containing acid solutions.

FINISHING PROCESSES		
Splitting and Shaving If the hide is thick enough it is split by a band-knife or if too rough it may be shaved in a cutting machine.	**Drying** The leather is dried under carefully controlled conditions.	**Fat-liquoring** Oils and greases are applied to restore softness.
Final Finishing Various processes to produce leather required for different functions.	**Dying** Color is added to leather either by dying or applying pigment to the surface.	**Currying and Stuffing** Further application of oils and greases in some heavy leathers.

footwear, belts, sports equipment, harness and saddles—but seldom is more than 3 percent of the total market value of an animal in its hide. Pig skins, used for footwear and clothing, are lighter than cattle hides but heavier than sheep and goat skins. Sheepskin leather is less common than sheepskin pelts from which the wool has not been removed (see 69–70).

Blemishes on hides because of bruises or insects or other damage often need to be removed in processing. Too many damaged areas may make a hide useless. Curing prevents putrefaction. The main methods are air-drying, salt curing and pickling.

Furs. Mink produce the most luxurious fur products, especially coats. They are essentially wild animals in captivity, kept entirely caged—generally individually, although litters may be raised together. Their diet of meat and fish by-products is usually mixed and fed daily. Special dry diets have also been developed. Feed and labor costs must be kept low by good management as net returns may be low in spite of the high value of the end product. A great variety of pelt colors have been bred, and pelts must be carefully matched and processed to produce the mink coats so attractive to consumers.

Furs are used from many other wild animals but the nutria (as the coypu is known in the fur trade), fox, chinchilla and beaver are kept most commonly. Foxes are sometimes kept on islands without other confinement. Where feasible, natural conditions are imitated in fenced enclosures. Rearing animals solely for pelts is not profitable unless feed and labor cost can be kept low.

Most rabbit skins come from animals slaughtered for meat. Farmers are encouraged to use white-furred breeds because the garments made from the skins can then be dyed a uniform brown color. The rabbit farmer cannot expect to obtain much of his income from fur since the price paid is very dependent on the fashion market—and in some seasons the skins have to be stored because there is no demand.

Feathers. Like hides, feathers are very much a by-product. Chicken and turkey feathers are cooked under pressurized steam to produce hydrolyzed feather meal. This meal, with about 85 percent protein, is used in rations for poultry, swine and other animals. The feathers may also be used in bedding and for ornamental materials. Duck feathers may be used as down for bedding and clothing but may be less satisfactory than goose feathers and down, which are highly prized. Goose down is plucked from live birds and is of the highest quality. Only ostriches are raised primarily for their feathers (see pp24–25).   CET

Ethics and Animal Husbandry

In the past 30 years efficiency in converting plant foodstuffs into animal products has brought down prices of the products to consumers, but it has required disturbing changes in farming methods.

Animals are fed diets and drugs which promote fast growth and early reproduction. The associated metabolic changes and the fact that many animals are killed at the end of their fastest growth period result in a shorter life span. There has been a substantial increase in the number of poultry, pigs, calves and even adult cattle kept indoors. Since buildings are expensive, the farmer needs to house animals at a high density. They are kept individually in pens and cages, or tethered in rows or crowded in groups, with frequent physical contact. Some do not survive well in these conditions, and the lives of all are very different from those of animals kept in fields.

Most people were unaware of the changes in farm animal management until these were brought to their attention by small groups concerned about the morality of these practices. Many farmers felt some concern about the modern methods, but they did cut costs and hence were necessary if the farmer was to be competitive. Those not economically involved with farming could stand back and ask whether the reduced consumer prices and, sometimes, increased farmers' profits warranted the treatment to which the animals were subjected.

What indeed are the rights of these animals? One of the first people to pose this question publicly was Ruth Harrison, whose book *Animal Machines*, published in 1964, aroused the interest and indignation of many. In Britain, letters were written to newspapers and to Members of Parliament about farming practices and about the effects on farm staff of regularly inflicting pain or discomfort on animals. The British government responded by setting up an advisory committee which produced the Brambell Report in 1965. This included such recommendations as to allow animals basic freedoms, for example the freedom of movement. Some of these were taken into account in legislation and the resulting welfare codes of the ministry of agriculture.

Public concern about farm animal welfare was considerable in some other European countries such as Sweden and West Germany but was less pronounced in

southern Europe or in important farming countries such as the United States or Australia. Hence there has been legislation or government advice on welfare matters in some countries but not in others. In 1976, the European Convention on the Protection of Animals kept for Farming Purposes required that animals be inspected daily and cared for, fed, watered, housed and given freedom of space and movement "appropriate to their physical and ethological needs." This was signed by 21 countries but has been ratified by only 12. Legislation in West Germany and welfare codes in Britain included reference to the Convention's requirements but such factors are not considered at all in many countries.

The issues which are of most concern at present are small battery cages for poultry, tethering and the use of small pens for calves and pigs, and various operations such as debeaking of poultry, tail-docking of piglets and lambs and castration. Is a battery cage whose floor measures 12×16 in

▶ **The hens that produce the supermarket eggs** are usually housed in small battery cages, often in dim light. The system pictured here involves manual feeding and egg collection, but in many large commercial units these are automatic. Sickness and injuries inflicted by pecking in the crowded conditions may go undetected for several days.

▼ **Small-scale egg farms** with free-ranging hens allow their birds much freedom of movement and the hens supplement their feed with nutrients that they find for themselves in the farmyard. It is argued that this gives their eggs a more natural appearance and flavor, and even makes them more nutritious, but none of these supposed advantages is proven (see p125). Some consumers are willing to seek out free range eggs, and pay a high price for them, simply out of disapproval for intensive methods.

(30 × 40cm) adequate accommodation for three laying hens?

Some people disagree with the consumption of any meat. Others believe that open-field systems for farm animals are better than indoor housing, a view not shared by those concerned about sheep freezing on hillsides in winter or starving on Australian rangeland during the dry season. The "welfare lobby" is not a uniform group for it includes extremists who would release all captive animals as well as people who are moderate in their demands and actions.

The questions raised in discussions of welfare are partly ethical, but concrete evidence about the responses of animals to particular treatments or housing conditions can be provided by physiological and behavioral monitoring. Animals in adverse conditions use several methods to try to cope with that adversity. The adrenal gland makes energy reserves available. Stereotyped behavior such as bar-biting may occur and morphine-like analgesics may be produced in the brain. The result of the behavior and of the chemical response is some alleviation of the adverse effects.

If the attempt at coping, using one or more of these responses, is unsuccessful, the animal may show reductions in disease resistance, growth rate and reproductive potential. It may also show abnormal behavior such as sucking the pen or damaging other individuals, for example by pecking out feathers in poultry or biting the tail region in pigs. The desirability of a rearing condition may be judged by the frequency and extent of such responses.

A different approach to assessing farm animals' welfare involves allowing animals to choose the characteristics of their surroundings. Such experiments demonstrate that most hens prefer a larger to a smaller cage and pigs prefer a pen with straw to one without. In order to interpret these results, however, the importance to the animal of the choice must be assessed.

Improved farm animal welfare can be achieved only if consumers accept higher prices and farmers accept lower profits. Such changes require government action, in the form of legislation or strongly worded advisory codes. Discussion between those concerned about farm animal welfare, farmers, scientists and government representatives is essential for change to be brought about effectively. It is improbable that a vegetarian will ever be satisfied by laws or agreed policies, but changes which are acceptable to the majority of the public are likely within a few years.   DMB

Ostrich Farming

The rise, fall and modernization of an industry

At the end of the 19th century ostrich farming had a spectacular rapid rise, and early in the 20th century an equally spectacular crash. The captive ostrich population in South Africa grew from zero in 1860, to 30,000 by 1875, to 750,000 by 1913, and had fallen to 23,000 by 1930. Today ostrich farming continues, as a small, thriving and fascinating industry.

In the 19th century the demand for ostrich plumes, which are about 2–3ft (60–90cm) long, for use in fashion soared with the realization that they could be obtained in large quantities not by hunting large birds but by breeding birds in captivity and plucking them repeatedly. A multi-million dollar industry sprang up. Huge fortunes were made as ostrich farms were set up in South Africa and later imitated in Egypt, Algeria, East Africa, Australia and the USA. The soft, white ostrich plumes were enormously sought after for adorning wealthy ladies in many parts of the world. Careful selection of good-quality birds was achieved, several slightly different and jealously guarded strains were developed. North African and Syrian blood was introduced by importation, and artificial incubators were developed. The industry crashed through a combination of war, recession, change in fashions and the arrival of the motor car (too low inside for hats with feathers, too windy for them outside). Many farmers were ruined.

Today ostrich farming is centered on Oudtshoorn in the Little Karoo region of Cape Province in South Africa. Some 80,000–100,000 birds are kept there, by about 300–400 farmers. The latter have learnt from the past, and ostrich farming is in most cases only a subsidiary activity. It complements conventional arable and stock farming, the growing of alfalfa (lucerne) for seed and for ostrich food, and the keeping of bees which feed on the alfalfa flowers.

Ostriches are kept in two main ways at Oudtshoorn. The best breeding birds are held as pairs in enclosures of 0.5–2 acres (0.2–0.8ha) which are planted with alfalfa and can usually be irrigated. In these the birds nest under artificial shelters and are usually allowed to accumulate and incubate their own eggs. The alternative system is to keep large numbers of birds of both sexes running free in paddocks several acres in area. There, with the large numbers of hens laying, the clutches become huge and the nests shambolic, so the farmers collect the eggs every couple of days and incubate them artificially. The hatching rate in the incubators is worse than that under nesting pairs

of ostriches. It is not clear how much this difference is due to the incubation method and how much to the fact that the eggs come from lower quality and less well-fed birds, and have been shaken about during collection and transport to the incubators.

Artificial incubation methods are still relatively primitive. After 38–44 days at 97–99°F (36–27°C), 50–80 percent of the eggs hatch, the chick often being helped out of the shell by hand. Experience has shown that the best way of rearing these chicks is to give them back to a breeding pair of ostriches. As happens in the wild, the adult birds accept and guard jealously many chicks apart from those they have hatched themselves. They are such good foster parents that for weeks they will put up with the farmers' removing older chicks and adding younger ones, to the extent that they care for a shifting population numbering, at any one time, up to about 80 chicks.

The young birds grow fast on a diet of growing alfalfa, corn and pellets. When two years old they are sorted and the best are selected for breeding, mates being chosen partly on the basis of their feather characteristics. Breeding birds may be kept for over 30 years. Of the remaining birds, the better individuals are kept for about 15 years to provide feathers. At 8- or 9-month intervals each bird is caught by cornering it and getting a walking stick round its neck. A hood is placed over its head to keep it quiet while

it stands in a small V-shaped crush pen. All its primary feathers are pulled or cut out, depending on age, and some of the body feathers too; enough are left on the back to prevent sunburn. The feathers are sorted according to size, color and quality, and made into articles such as hats, feather boas, dusters and various items of knickknackery sold to tourists.

Whereas in the past the feathers were the only important harvest of the ostrich industry, now they are only one element. The majority of ostriches are slaughtered at maturity and virtually all parts of them are used. The skin goes to make a strong but

▲ Pastoral scene in South Africa. Ostrich farming has been tried in Algeria, East Africa, Australia and the United States, but since 1914, when the fashion market for ostrich feathers collapsed, only in South Africa have these birds been successfully farmed.

◄ Plucking pen. Ostrich plumes taken from the wings were the original high-fashion product leading to the bird's domestication. Today, feathers from other parts, used for feather dusters, are just as important. By-products include skins, meat, bonemeal and entertainment for tourists.

supple leather for handbags and purses. Meat from the thighs (ostrichs contain very little meat elsewhere) is dried and sold mainly as biltong (strips of meat) as a delicacy. The rest of the carcass is converted into meat or bonemeal.

Around Oudtshoorn a tourist industry has grown up and provides another source of revenue for the few large ostrich farms that open their gates to visitors. Tourists pay for guided tours around the farms, watch a short ostrich race, perhaps even sit on and probably fall off an ostrich themselves, and buy a range of ostrich-related items in curio shops. Decorated eggshells, ostrich-foot

table lamps, feather handbags, eggshell jewel boxes, and ostrich-skin purses are among the items popularly bought by the tourists.

Ostrich products bring in about 7 million US dollars per year. About 40 percent of this income comes from feathers; on slaughter, about four-fifths of the value of the carcass is in the skin. The income from the related tourist industry is much harder to assess, but is appreciable. It has been an extraordinary development over a period of domestication of only 120 years, only two or three lifespans of the world's largest bird.   BCRB

THE PRODUCERS

Numerically, by far the commonest farmed animal is the domestic fowl but, by weight of products, birds are less important than mammals. Poultry numbers have increased considerably in recent years and several countries are expanding their poultry industries so the chicken's exploitation by man is likely to expand further. The mammals which do best on man's farms are cattle, sheep, goats and pigs. Cattle are particularly well adapted to survive and reproduce in most environments imposed by man and are now commoner than sheep. In the damp tropics the buffalo does well and in arid areas, camels, goats and some sheep breeds are successful. When the pasture is short or sparse, or the temperature low, sheep are often the most valuable farmed mammal.

Pigs provide only meat but are a very important source of it in the wealthier countries. Rabbits are also increasing in importance.

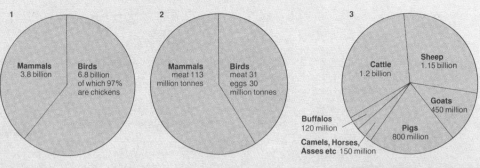

1

Mammals
3.8 billion

Birds
6.8 billion
of which 97%
are chickens

2

Mammals
meat 113
million tonnes

Birds
meat 31
eggs 30
million tonnes

3

Cattle
1.2 billion

Sheep
1.15 billion

Goats
450 million

Buffalos
120 million

Camels, Horses,
Asses etc 150 million

Pigs
800 million

◄ **More birds are farmed (1) but mammal
products are by weight (million tons per
annum) more important (2). Among the most
common farmed mammals (3), cattle and sheep
predominate.**

▼ **Mass production of turkeys in Wisconsin,
USA.**

BREEDING AND GENETICS OF FARMED ANIMALS

▶ **Thriving on sparse forage,** this "cattalo" results from crossing bisons and Herefords. As in crosses between zebu cattle and yaks or mithuns, the hybrid male progeny are sterile or have very low fertility.

▼▶ **Wide original distributions** of the species which became the most common farmed animals have contributed to genetic diversity. Crossbreeding geographically isolated types to develop new combinations of useful traits has thus been an important breeding strategy.

MAN has selected particular groups and individuals for breeding from among his domesticated animals since the ancient civilizations of Egypt, Greece and Rome. This has been done to develop animals better suited to particular purposes.

Strains domesticated by prehistoric man decreased in size compared with the wild forms from which they originated. But with the advent of conscious animal breeding, diversity of type was developed and size was increased again. Thus the ancient Egyptians had definite breeds of dogs, sheep and cattle; the Greeks produced particularly fine-wooled sheep as they developed the skills of breeding; and under the Romans, who further improved the knowledge of breeding they derived from the Greeks, cattle were developed to a size some 25 percent taller than those of the earlier Iron Age.

With the decline of the Roman Empire purposeful breeding decreased and a number of breeds were lost. Livestock again declined in size as the animals were no longer bred for specific purposes. This trend was then only halted by the revival of interest in classical animal breeding in the 15th and 16th centuries.

In addition to the influence of man in shaping his domestic animals there has also been the influence of nature. The effects of environmental temperature and climatic conditions, of feed availability and quality, and of parasites, diseases and predators have largely determined the way a population of domestic animals has developed in any particular environment. The adaptation of the dromedary (Arabian camel) to desert environments or of the River buffalo to hot riverine conditions are obvious examples. But also important are the more subtle adaptations, for instance in the regionalization of some sheep breeds, and the variations within a breed. Across South Africa and Australia increasing dryness of climate and scarcity of feed are reflected in increasing coarseness in the fleeces of the sheep.

Influences of nature were dominant from the end of Roman times to the 18th century and beyond. By the 18th century in Europe, and for a considerable time before, farm livestock were usually known by the name of the region whose environment shaped them. Regionalized farm livestock tended to be multipurpose animals; the same breed of cattle, for example, would be used for draft, meat, milk and hides. There was throughout the century, though, a growing interest in agricultural experimentation.

In the United Kingdom there was a number of improving breeders of whom Robert Bakewell (1726–95) is the best known. He carefully selected and crossed both cattle and sheep to produce quick maturing animals suited to the requirements of the butchers, fixing his desired crosses by inbreeding. In France and Austria, Spanish sheep were introduced to improve the fineness of the local wool. Increasingly, wealthy farmers and landowners were prepared to travel long distances to obtain the best animals to use as breeding stock. The close adaptation to local environment of farm livestock was in this way broken down, and the purpose for which an animal was bred became more specific.

These developments came to be influenced by the agricultural shows. Originally only the final products, the fat stock, were exhibited and awarded prizes at these shows. But by the middle of the 1820s prizes were being offered for store stock (animals bought or held for fattening) and breeding stock as well. Initially the normal environ-

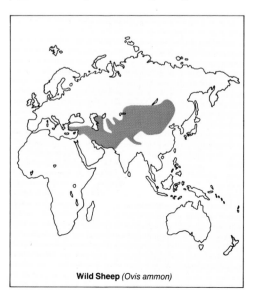

Wild Sheep (*Ovis ammon*)

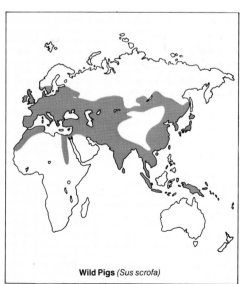

Wild Pigs (*Sus scrofa*)

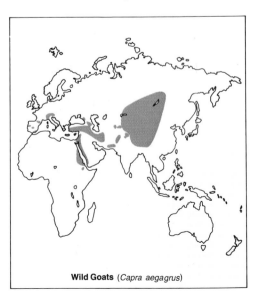

Wild Goats (*Capra aegagrus*)

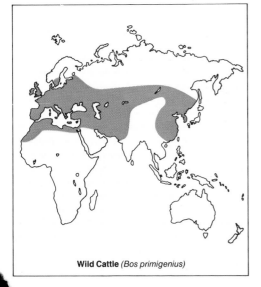

Wild Cattle (*Bos primigenius*)

ment of the livestock and its effect on productivity, was taken into account in establishing show classes, but in the second half of the century attention in the show ring came to be focused solely on breed. Show judging, with few exceptions, was based on visual assessment, and as the show classes for livestock were redefined to include not only one breed, the visual characters looked for were more carefully specified and increased in number.

Breed societies were established which laid down the points that animals of the breed should conform to, and much attention was also given to the pedigree, or parentage, of the animals. Pedigrees of Shorthorn cattle (see p40), were first published in 1822 and in later years herd books

were established for other breeds by breed societies. (Herd books and flock books are kept by the breed societies to record the pedigrees of the registered animals of the breed.) The Shorthorn breeders set the fashion in showring and herd-book practices and by the 1870s breeders from as far away as the United States and Australia, who lacked their own pure breeding stock, were competing at British auctions to purchase prize-winning Shorthorns of fashionable pedigree, and inflated prices were paid for these animals. These trends fixed the characteristics of particular breeds, in visual terms, much more precisely than before. Showring competitions did much to foster livestock breeding but they have also been criticized for lack of relevance to the wider

farming industry. Many of the visual criteria, such as particular color markings, on which animals came to be judged were of little relevance in commercial farming.

The beginning of the 20th century saw the rediscovery of Mendelian genetics and a steady increase in the application of genetic principles to animal breeding. Theory tended to confirm what many breeders were already practicing but gave the breeders a much more precise understanding of the likely outcome of their actions. Genetics has been exploited to the greatest extent in poultry production.

With poultry, the interval from one generation to the next is short compared with large farm animals. Population size of a breeding stock can be large, thus increasing the opportunity for selecting particular individuals, and environmental and nutritional conditions can be closely controlled. Large international firms have grown up which specialize in producing day-old chicks which are then sent to the farmers. Their breeding programs take no account of the old poultry-breed criteria of the showring. Instead the genetic potentials of various lines of birds are determined by quantitative

techniques. Crossing programs are carefully calculated to produce the desired gene frequencies in a population. Resulting parent lines are then crossed to give hybrid offspring, showing hybrid vigor—the day-old chicks sold to the industry. The hybrid birds are judged solely on measurable criteria of commercial importance, such as rate of food conversion, or egg production. Similar methods have also been applied to a lesser extent in pig breeding.

Cattle and sheep breeding have been influenced more, in the present century, by the simple expedient of replacing visual by quantitative assessment. Selection of animals for breeding has been based more on such factors as measured milk production in dairy cattle, on weaning weight for meat animals or on wool weight for sheep.

With the introduction of quantitative measurement, animal scientists have been able to provide better guidance to breeders; for example by establishing estimates of the hereditability of various economically important characteristics. Visual factors are still taken into account particularly where they are likely to affect production or value.

In many parts of the world it is difficult

▲ Charolais bull with Friesian cow and calf. Often only the best dairy cows are bred with dairy bulls. Poorer producers may be bred with beef bulls to give calves reared for meat.

► Mendelian genetics in sheep. The 19th-century Austrian monk Gregor Mendel discovered the principle leading to today's understanding of dominant and recessive traits in inheritance. A trait such as hair color or the presence or absence of horns is controlled by a pair of genes, each inherited from a different parent. Genes may have two or more alleles (alternative forms). If both of an individual's genes for a trait are the same allele, then the individual is "homozygous" for that trait, eg the cattle and sheep at (A). If the two genes are different alleles, then the individual is "heterozygous," eg the offspring at (B). In a heterozygous individual, one of the alleles is usually "dominant" (represented by uppercase letters in this diagram), the other "recessive" (lowercase letters), and the heterozygous individual displays the dominant trait. If it mates with another individual carrying the recessive allele (C), some of its offspring may display the recessive trait. Genes are segments of paired chromosomes, and those on the same chromosome are linked in reproduction. Rules for the inheritance of a single trait, or of linked traits, can be set out relatively simply, but as animals have several chromosomes (sheep have 27 pairs), large numbers of traits are not linked, and the combinations are complex.

Principle of Mendelian genetics when considering one character type.
A gene can occur in alternative forms or alleles. For example in sheep the gene for coat color gives a white or black fleece.

White sheep with the white allele only. A dominant characteristic.

Black sheep with the black allele only. A recessive characteristic.

A. If **WW** sheep are crossed all offspring will be **WW**.

If **ww** sheep are crossed all offspring will be **ww**.

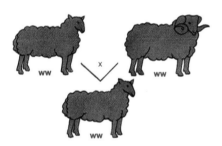

B. If **WW** sheep are crossed with **ww** sheep all offspring will be **Ww**.

Ww offspring will be white in appearance

C. If **Ww** sheep are crossed the ratio of character types will be 1 true white, 1 true black and 2 white sheep.

WW: True white sheep, only producing true white offspring if crossed together.
ww True black sheep, only producing true black offspring if crossed together.
Ww White sheep, producing black and white offspring if crossed together.

Principle of Mendelian genetics when considering two character types.
With two pairs of alleles on separate chromosomes the situation becomes more complicated. For example, when crossing Angus and Hereford cattle.

Angus cattle: Black coat with no horns. Both characteristics are dominant.

Hereford cattle: Red coat, and horns. Both characteristics are recessive.

A. If **BBPP** cattle are crossed all offspring will be **BBPP**.

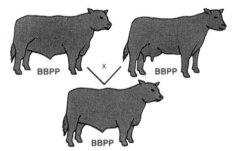

If **bbpp** cattle are crossed all offspring will be **bbpp**.

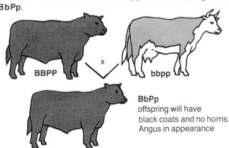

B. If **BBPP** cattle are crossed with **bbpp** cattle all offspring will be **BbPp**.

BbPp offspring will have black coats and no horns, Angus in appearance

C. If **BbPp** cattle are crossed the ratio of observed character types will be 9 black without horns, 3 red without horns, 3 black with horns, 1 red with horns. Only one animal will be pure Angus and one pure Hereford.

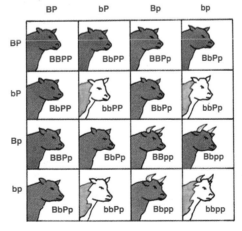

	BP	bP	Bp	bp
BP	BBPP	BbPP	BBPp	BbPp
bP	BbPP	bbPP	BbPp	bbPp
Bp	BBPp	BbPp	BBpp	Bbpp
bp	BbPp	bbPp	Bbpp	bbpp

to achieve very rapid progress in breeding the larger farm animals. Many farmers do not have large numbers of animals available to them from which to select, although in dairying this is sometimes overcome by the operation of nationally organized breeding schemes involving performance testing of bulls and the use of artificial insemination. Environmental conditions can seriously affect the performance of livestock under field conditions. European cattle and zebu cattle (see pp44–47) can be crossed to give varying proportions of zebu blood If the animals are grazed together in a tropical environment, in a good year with plenty of feed and low heat and low disease stress, the animals with a low proportion of zebu blood will grow best. In a poor year with little food available and high heat and high disease stress the animals with a high proportion of zebu blood will grow best. It is important to define carefully not only economic characters but also the conditions under which the animals are to be reared.

There is a continuing need for a range of livestock types to suit various conditions of production. There is also the need to maintain a range of types for future breeding purposes so that changing market needs can be met. Consumers now prefer meat with much less fat than in the past and breeding programs have had to be changed to meet this preference.

In spite of this, producers in highly industrialized countries have been encouraged to use fewer and fewer breeds of livestock. Highly controlled conditions of production and a tendency to stress one market requirement above all others have created this trend. There is now concern that much of the genetic variation developed over many centuries in different environments will be lost. If this happens the ability of future generations of animal breeders to adapt their livestock to changing market requirements or changing production conditions will be seriously curtailed.

Organizations such as the Rare Breeds Survival Trust in England have been established to preserve breeds of livestock no longer favored by farmers for commercial production. Some breeds have become extinct in this century but others, though low in numbers, have now been assured of survival. England has led in this field and the Food and Agricultural Organization of the United Nations is also taking an interest in this work, so that the examples of the wide diversity of genetic material that exists in the farm animals of the less developed world will be preserved.   LJP

CATTLE

Bos primigenius ("B. taurus," "B. indicus")
One of five species of the genus *Bos*.
Order: Artiodactyla.
Suborder: Ruminantia.
Family: Bovidae.
Number of breeds: approximately 200.
Distribution: worldwide between 60°N and 40°S; humped zebu cattle mainly in tropical climates and humpless cattle in others. Sanga cattle (zebu crossed with humpless cattle) mainly in Africa. Small feral populations in many places, most notably Chillingham Park, N England.

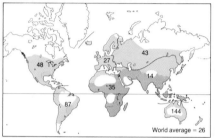

Figures show number of cattle, by region, relative to 100 head of human population (FAO data 1982).

European	Zebu
Intermediate	Zebu recently introduced

Size: height at withers up to 72in (180cm), weight 1,000–4,000lb (450–1,800kg).

Horns: absent (polled), or short to long; cylindrical in shape, set wide apart, curving; vary strikingly with breed.

Other features: long nasal bone and flat forehead with eyes well below horns; back straight, hair short, back vertebrae sloping towards the rear; 13 pairs of ribs; zebu and sanga cattle have a hump of muscle and fat, not resulting, as in other humped Bovini, from long processes on the vertebrae.

Diet: grass, weeds, browse.

Dental formula: 10/4, C0/0, P3/3, M3/3 = 32. (Note premolars are molar-like.)

Longevity: 5–40 years.

Breeding: nonseasonal; gestation about 283 days; twins rare.

► **With a face that denies him breed status,** this almost Black Welsh bull stands a wary guard in his native terrain. Descended from cattle the ancient Britons took with them as they retreated from Saxon invaders, black Welsh cattle were traditionally dual-purpose milk and beef animals. Today, the registered Black Welsh is mainly a beef producer. Cows can weigh 1,780lb (810kg), and bulls weigh up to 2,464lb (1,118kg).

THE 1,200 million cattle on this planet are located worldwide. About 46 million tons of beef and about 480 million tons of milk are produced annually—enough milk to provide each man, woman and child with two-thirds of a pint of milk daily. The host of local types and breeds of cattle may be divided into two major groups, the humpless cattle of European and northern Asiatic origin and the humped zebu cattle which trace their origins and popularity to India.

Humpless cattle are noted for their placid temperament and capacity for meat production. They are widespread in Europe, North and South America, Japan, Australia and New Zealand. Heat and insect resistant, the humped cattle are predominant in southern and Southeast Asia, the Middle East, Central Africa, Brazil and along the Gulf of Mexico.

The two groups have been regarded by some authorities as separate species, with the name *Bos taurus* for the **European type** and the name *B. indicus* for **zebu cattle**. By modern scientific convention, however, they are counted as only one species, with the same name as their ancestor *B. primigenius*, although the names *B. taurus* and *B. indicus* remain in common usage.

Other farmed members of the genus *Bos* are the **yak** (*B. mutus*) of Tibet (see pp62–63) and the **mithun** in India (see p49), a domesticated form of gaur (*B. gaurus*). **Bali cattle** (see p48–49) are a variety of banteng (*B. javanicus*), a wild animal of Indonesia. The kouprey (*B. sauveli*) has never been domesticated. It became known to zoologists only in 1937, in Indochina, and may well now be extinct. Together with the Asiatic buffalos (genus *Bubalus*—see pp 58–61). African buffalos (*Synceros caffer*) and American and European bison (*Bison*), the *Bos* species make up the tribe Bovini, all termed cattle.

Origins of Cattle

Fossil wild cattle remains have been found in deposits from the lower Pleistocene (about 2 million years ago) in the Siwalik Hills of northern India. By the end of the Pleistocene (11,000 years ago) the humpless form was especially widespread, occupying wide reaches of Asia, Europe and North Africa.

Earliest man appears to have held wild cattle, above all other animals, in fear and awe, and to have worshipped them as gods. At the same time he viewed the bull hunt as the ultimate test of his skill, ingenuity and manhood. Cattle drawings have been found at Lascaux in France dating from 15,000BC. One painting shows a large red dappled cow

with long horns directed upward and outward. Human beings are close by. No weapons are in their hands and the cattle are standing quietly. This may indicate an early date for domestication in Europe, but most authorities estimate that cattle were first domesticated about 9,000 years ago in the Middle East.

Domestication was probably achieved by capturing orphaned calves. To have a degree of control over the maturing bull calf, early man used nose-ringing, dehorning and castration. Even today the "domesticated" bull cannot be easily handled, and in certain cultures bull-baiting games are still a way of proving manhood (see "Fighting Cattle," p38).

The original selection of breeding animals which were to spawn the prime milk and meat producer of the world had little bearing on productivity. Cattle were originally selected for their crescent-shaped horns, in honor of the Lunar Mother Goddess, worshipped over an immense area of the ancient world. The moon's regularly changing phases made it a fertility symbol, and its crescent shape made the horns of cattle magical.

In such cultures oxen were first a source of animal power. Cows and bulls had to be tame enough to be kept in the temple corral because they were used to pull carts in the temple processions and in sacrificial rites. The requirements for a domesticated animal suitable for religious purposes were a quiet disposition, strength and the appropriate horn shape. Much later, milk, meat and leather became the main concerns.

In the early domestication of cattle, humans learned to change their body characteristics by controlled breeding. Inbreeding reduced the size of tame cattle, but by breeding tame cows to wild bulls, larger calves were obtained.

Almost all great civilizations were built by people with a bull and a cow culture. Cattle people gained ascendancy over sheep or goat people in the Middle East and the sheep-based culture became subordinate. Cattle cultures and grain production went together, and in such settled communities learning, knowledge and civilization first took root. Mongolia's horse-culture had to dominate a more highly civilized cow and pig culture when they conquered China.

In India a major ancient cow culture has survived. India ranks first in numbers with 200 million of the world's cattle, and they continue to be revered. The early priests designated the cow by a word which means "mother-cow." All but two Indian states

World Importance of Cattle

The world's estimated 1.2 billion cattle are found mainly between 60°N–40°S. Slightly more than half are in Europe, the Western Hemisphere, New Zealand and Australia, where humpless cattle predominate, producing milk and beef.

Some are specialized dairy or specialized beef breeds, others are mixed-use breeds supplying both of these products. Formerly work was a product of cattle in these areas of the world but now tractors have taken over their role.

Slightly fewer than 40 percent of cattle are in Asia. Here humped zebu breeds are most common. They are primarily work animals but often supply milk as well. Usually, cattle are slaughtered for beef only when they are no longer useful, and then only if there is no religious prohibition against it. Chinese yellow cattle and the African sanga breeds are the result of crossbreeding between zebu and humpless cattle in the remote past.

Slightly more than 14 percent of cattle are in Africa, mainly zebus and sangas, but humpless forms are important in North and West Africa. About 80 percent of African cattle are in the hands of indigenous Africans. The large majority, held as a matter of prestige, are milked but used only meagerly for work and meat. As cash economies encroach on traditional ways of life, many are being sold for slaughter.

About 40 million tons of beef and about 400 million tons of milk are produced annually in the world—enough milk to provide each man, woman and child with 10 fl oz (300ml) daily.

Number of cattle in the world (millions)

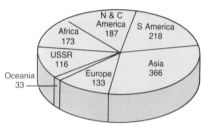

N & C America 187
S America 218
Africa 173
USSR 116
Europe 133
Asia 366
Oceania 33

World total: 1,226 million

ANCIENT NORTH AFRICANS AND THEIR CATTLE

▶ **In the central Sahara,** one of the many thousands of rock paintings left by cattle-herding peoples who occupied the Tassili n'Ajjer in present-day Algeria from 4000–1200BC. In their gradual migration to Europe and West Africa, both short- and long-horned varieties of cattle passed through the hands of the Tassili herders, who enjoyed a much milder climate than the desert conditions of today.

▼ **Threshing grain in the tomb of Menna** in the ancient Egyptian city of Thebes. Farmed cattle first appeared in Egypt around 4500BC, derived from long-horned cattle domesticated earlier in the Middle East. Today, short-horned breeds are the most common in Egypt.

have enacted laws strictly forbidding mistreatment or slaughter. Special cow nursing homes called Gowshallas have been established to take care of barren animals and those unable to work. Twice a year, in late summer following the rains and following the harvest, there are cow festivals throughout the country. Cows are decorated with garlands of flowers, fed succulent forages and sprinkled with holy water.

The best-known Indian center of domestication was in the Indus valley, which had early trading links with Mesopotamia and Egypt, both centers of domestication as well. A temple frieze dating from about 3000BC in the Mesopotamian city of Ur shows imported Indian humped cattle. Humped varieties were being bred alongside humpless ones by 2000BC in the Indus civilization, where crossbreeding must have occurred.

Egyptian wall paintings give prominent place to a now-extinct breed of humpless longhorn, the Hamitic. From about 2000BC it began to be crossbred with introduced zebu stock to produce the ancestors of Africa's widespread sanga breeds (see pp47–49). West African N'Dama cattle (see p38) may be a purer descendant of the Hamitic Longhorn.

Egyptian wall paintings also show a short-horned ancestor with the high forehead it shares with its purest descendants. It may have been domesticated in the Middle East before its man-made migration first to Egypt, then through North Africa to Spain, France, Switzerland and the Channel Islands, leaving a trail of shorthorn breeds: Galacian Blond, Blonde D'Aquitaine, Aubrac, Tarentaise, Brown Swiss and Jersey. The ancestral animal existed in the late Stone Age in Europe, and remains have been found in Britain as late as the time of the Roman occupation. They have usually been found in or near human habitations.

The long-horned humpless ancestor in its wild European form is known as the aurochs. Extinct in 1627 when the last cow died in Poland, the bulls of this variety sometimes stood almost 6ft (2m) high at the shoulder. Lyre-shaped horns have been found which measure almost 3ft (1m) from tip to tip. They are said to have been extremely large, strong, fast, clever and brave beasts. The bulls were virile, with impressive harems.

In prehistoric times domesticated long-horned cattle were taken into Europe directly from the Middle East through Turkey and the Balkans. There is little evidence of how much domestication there was of the native European aurochs. Cattle herding immigrants around Lake Zurich in Switzerland are thought to have used local wild cattle in breeding new strains. The aurochs was also being domesticated in Denmark and the Schleswig-Holstein area of Germany around 3000–1500BC. The long-horned European breeds of today are a

mixture of Middle Eastern, African and European strains, the detailed origins of any particular breed bordering on mystery. British White Park cattle (see p37), for example, resemble white bulls which ancient druids used in their sacrifices, but they are also like Italian Chianina cattle (see p43). They could be descended from herds abandoned by the Roman legions when they withdrew from Britain.

Distribution of types in the tropical and subtropical portions of the world reflect natural migrations of men and cattle along overland routes. Overseas migrations of men and available cattle from their native lands have continued into historic times. The first of the Spanish Criollo stock (see p38), from which the Texas Longhorn derives, were landed in the New World in the 15th century, and almost every major European breed has found its way into the Western Hemisphere since. British cattle first appeared in Australia at the end of the 18th century. Zebu cattle first appeared outside Asia and Africa in the mid-19th century. American Brahman cattle (see pp46–47) are a New World zebu breed. In the 20th century cattle have been moved freely by sea and air, and now frozen semen and fertilized eggs can be transferred easily to any country where the techniques of artificial insemination and artificial implantation of eggs are practiced.

Human intervention in cattle migration and breeding has not been confined to the multiplicity of domesticated cattle breeds which have resulted. Yaks have long been crossed with zebu and recently with Highland cattle. Bantengs and mithuns are also traditionally interbred with zebu cattle. Hybrid males are normally infertile, but the modern age presents new possibilities. In the 1960s a fertile "beefalo" was developed in California which is claimed to be three-eighths American bison, three-eighths Charolais and one-quarter Hereford.

Cattle Biology

Humpless and zebu cattle differ from other bovines by the cylindrical shape and the position of their horns, set wide apart on a ridge on top of the skull. In zebu breeds the horns arise from the sides of the skull and grow upwards or slightly back. In humpless forms they arise where the back and side margins of the skull meet, growing out from the sides of the head before curving, usually first forward then upward. Polled (hornless) cattle are the result of selected breeding.

All cattle share a distinctive arrangement of teeth. The upper jaw is toothless except for molars, and eight incisors in the lower jaw bite against a dental pad on the roof of the mouth. When grazing the animals grasp a tuft of grass then lift their heads, shearing the tuft off.

Like many other hoofed mammals, cattle are ruminants. They have a complex stomach with four chambers including one where micro-organisms digest the cellulose which is so prevalent in their fibrous diet. The animals can digest the constantly multiplying micro-organisms themselves as they spill into other chambers, and also digest the fermented materials into which the micro-organisms break down the cellulose. Undigested cellulose is regurgitated to the mouth where the animal "chews a cud" before swallowing again.

Being ruminants, cattle can make use of nonprotein nitrogen in the form of urea as well as by-product feeds such as brewery waste, straw, beet pulp and almond hulls. They serve mankind best when they are on land where only coarse plants are available. The grazing choices of range plants by cattle are 70 percent grass, 20 percent weeds and 10 percent browse. Three major grazing phases of from 7–9 hours have been observed in cows for each 24-hour period, with resting and rumination periods between. Heavy grazing occurs near dawn, before and into dusk and around midnight.

All cattle are basically herd animals, irrespective of actual farming practice. In a free-ranging or semi-wild state, a herd has a "king" bull which is up to 40 percent larger than the cows and has much greater strength in his neck and shoulder muscles. Cattle mate at any season, but cows are rarely receptive during a calf's early stages of suckling. If a herd grazes widely over the range, a mating strategy develops whereby cows which come into heat begin riding one another. The bull can detect receptive cows when he sees distant mounting.

Cows withdraw from the herd to calve, wild cows totally isolating themselves from the group. The calf "lies out" away from the herd and its mother will attack intruders. A wild cow must seek "king" bull approval for herself and her calf before rejoining the herd. Domestic cows only partially isolate themselves. They will accept help from intruders, although beef cattle will defend their calves more readily than dairy cows. Twins are rare and the calf is capable of a high degree of independent activity from birth, standing from 30–60 minutes after birth. The mother goes to it 4–6 times a day, when it suckles. Lying-out and the mother's

head threats against curious relations both act to set greater distances between cows in a herd than between sheep in a flock. The young stay in the cow herd after weaning but males leave the group at about two years of age, join a male sub-group and fight for mating rights, most of which fall to the "king" bull.

Longevity ranges from a few years to a few decades. One dairy study found that 12 percent of all heifers that came into milk at least once reached five years of age. One dairy cow in 1,000 attains the age of 13 years, and only one in 100,000 passes her 19th birthday. The cattle longevity record appears to be a beef cow in Wales which died in 1956 at 40 years. An American Jersey-Holstein crossbred dairy cow died in Wisconsin in 1979 at 39 years.

Cattle Products

Overwhelmingly, the most important products from cattle are meat (see pp12–14) and milk (see 16–19), but there is more to be had from a cow, bull or steer than milk, steak or hamburger. In the not too distant past, thousands of cattle were raised and slaughtered on the pampas of Argentina and along the coastal plains of the Gulf of Mexico and California, solely for their hides and horns. Hides now account for about 3 percent of the value of slaughtered cattle and about half of by-product sales.

Horns have been used to make wearing apparel, in the form of walking sticks, combs, buttons and knife handles, to make music and to summon and warn people. Horns have also been used as inkholders by early scholars and writers, and to dispense liquid doses (drenches) to cattle, as well as to hold gunpowder.

An average market steer (1,000lb–450kg) yields about 440lb (200kg) of beef. About 40 percent of the animal's live weight

▲ **Winter combat.** A third of Chillingham bulls display permanent injuries inflicted by the horns of rivals.

◀ **Sacred cow.** Taking up the center of the main road in Port Blair, Andaman Islands, India, this cow commands too much respect to be roughly pushed on her way. The repugnance among Hindus towards eating meat derives partly from a belief that human souls can be reincarnated in the bodies of animals. The reverence for cattle in particular also reflects thousands of years of dependence on this species' labor and milk, as well as its ancient status as a symbol of fertility.

▶ **Bovine body language.** Differences in tail posture enable the animals in a herd to read each other's mood and condition. (1) A cow's tail hangs straight down when she is relaxed, grazing or walking. (2) When she is cold, sick or frightened, it tucks between her legs. (3) During mating, threat or investigation, the tail hangs away from her body. (4) When galloping she holds it almost straight out. (5) During bucking or gambolling it has a kink.

away from receptive females. The young bulls (maturity is reached at about 2 years of age) stay in the main herd with the cows and calves. This main herd roams the 330 acre (134ha) park freely and it seems that mating is by the dominant bull of the home range that a cow happens to be in when she comes into heat. Breeding takes place all year round.

The bulls are very vocal, calling and lowing in reply to each other, and frequently probe the boundaries of adjacent home ranges. They are more likely to display to their neighbors than to fight with them but very frequently have "fencing bouts" with the bulls with which they share the home range. Fighting is dangerous and at any time about a third of the bulls are seen to have sustained permanent damage like the loss of an eye or a punctured body wall, but fatalities are rare.

Social interactions are most frequent during summer. The cows interact with little formality or display, unlike the bulls who normally precede or follow a fencing bout with threats and assertions. Bulls and cows also differ in their patterns of grazing, ruminating and standing. Bulls graze and ruminate in shorter bouts and are more easily distracted. When they have nothing else to do, cows are likely to lie down, but bulls tend to remain standing (presumably to stay alert for rival bulls or receptive cows).

Cows have their first calf at the age of 3–5 years. Some cows may still not have their full adult dentition by the time they have their first calves. Bulls, however, probably have only a short reproductive life and their fertility could well be rather low. Their testes are small, even taking into account the small body size. Although they are capable of inseminating from the age of two years, it seems that they do not have the opportunity of doing so until they take up the "home range group" or "king bull" mode of living. Depending on the competition, few bulls above seven years of age achieve many matings.

Cows, however, may still be capable of bearing calves at the age of 15 years. The maximum life span of bulls is about 9 years and of cows about 17 years, although only a small minority reach such ages. Most winter-born calves die and there is heavy mortality in both sexes up to the age of four years.   SJGH

The Chillingham Wild White Cattle

How do cattle behave, free of man's interference? As the wild ancestor of our domestic cattle, the aurochs, is extinct we can only answer this question by studying cattle which have been allowed to revert to the wild state. The Chillingham herd is the most ancient such herd; it has inhabited Chillingham Park (northern England) probably since the park was created in the 13th century. We do not know where they originated, but they resemble medieval domestic cattle in general size and shape, and their uniform color (white with red-brown

ears) must have been the result of human selection. There are other ancient herds of white cattle in Britain but none has been kept pure or has so long a history.

The Chillingham cattle are never handled and no bull calves are castrated. Bulls of about four years of age and over live in groups of two or three which share home ranges (territories). These groupings and home ranges (each roughly 100 acres–40ha) last for life. Strong bulls do not expel the old and feeble bulls from their home ranges but they do assert their dominance and chase them

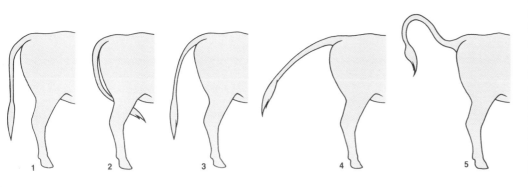

becomes by-products. At one time, leather upholstery was stuffed with animal hair, but synthetics have almost completely taken over this market. The hair is still used, however, to manufacture artists' paint brushes, although that market is limited because only the fine hair from the animal's ear can be used. (They are sold as "camel's-hair brushes".) Hair contains protein, and from it researchers have developed a feed additive for livestock—a less expensive alternative to grain as a protein source.

Tallows and greases are probably second to hides in terms of cash value for by-products. On average, a steer yields about 60lb (27kg), including both edible and inedible forms. Consumers have largely turned away from animal fats for cooking and shortening uses, but many margarines contain rendered animal fat; if a plant source is necessary for one's diet the label should be checked carefully. Today's major edible use of tallow and rendered fat is its addition to livestock and pet foods. Some inedible tallow is still used industrially for lubricants, although its bigger market—soap—has virtually disappeared with the introduction of synthetics. Concern over environmental pollution, particularly from detergents, has rekindled interest in natural-based soaps.

Variety meats are also on the by-product list, including liver, hearts, stomachs, tripe, spleens, kidneys, brains, tongues and sweetbreads. Gelatin from the steer's collagen or cartilage is used for salads, desserts, chewing gum, marshmallows and soluble medicine capsules. Inedible uses include photographic film, cosmetics, printer rollers, window shades, sheet rock, matches, wallpaper, sandpaper, glues and adhesives.

Beef glands also contribute products used in pharmaceuticals, including insulin. Tankage, bone meal and fertilizer come from the remains. Also, from the stomach of dead or unborn calves comes rennet which causes milk coagulation, an indispensable part of cheese making.   JLA

Long-horned Cattle

The Longhorns of Texas, the fighting bulls of Spain and the noble white cattle of the grand English parks are the most famous and romantic long-horned cattle. But others have played their part in history and many may have a future role to play. Some were selected by man to fulfill specific functions and others show characteristics probably molded by their environments.

Harsh winters favored the hardiness of the Highland cattle of Scotland. They are long-lived and produce excellent beef. These qualities (under-exploited in their native land where most hill grazing is used by sheep) have made them popular in many parts of the United States and Canada, where their slow rate of maturation can be overcome by extra feeding. The N'Dama cattle of West Africa are adapted, in contrast, to the hot tropics. In addition to a remarkable heat tolerance, better even than the zebu (see pp44–47), they are not as susceptible as many other cattle to the tsetse-fly transmitted trypanosomiasis

Fighting Cattle

The sport of bullfighting has flourished in Spain, Portugal and France during the last 100 years especially, but has roots in the ancient past, and a lively history in all ages. Each of these countries has its own variants, and the French and Spanish traditions are completely different.

In the Iberian Peninsula great toreros, like Manolete, are idolized like pop-stars; some, matadors, fight the bulls on foot, while rejoneadores work the bulls from the back of a superbly trained horse. These beautiful animals, usually of the showy Andalusian or Lusitana breeds, are trained to sidestep a charging bull with such precision that the rejoneador can plant a sword between the bull's shoulder blades.

In Portugal bullfighting is less gory than in Spain. The rejoneadores do not kill the bull and are followed by teams of a dozen young men clad in gay traditional costume who catch and hold the bull, which may weigh 1,320lb (600kg), with their bare hands. They then release it with one man holding the tail as the others escape over the barricades of the arena.

In Aquitaine (southwest France) young men somersault over the back of a charging cow. Their variant of the game (*course landaise*) is very similar to that of the Cretan bull-leapers 4,000 years ago.

Perhaps the most animated bullfights are

those of Provence (*course camarguaise*), which originated in the Camargue, the delta region of the Rhone River. Here black bulls carry tiny rosettes attached by strings to their horns. Young men in white (razeteurs) try to remove the rosettes and the strings with the help of a metal comb without being caught by the bull. Great bulls are lively, combative and run year after year. They are often better known (and loved) than the razeteurs and die of old age—a complete contrast with the short life of a Spanish fighting bull.   PD

which makes large areas of Africa uninhabitable by cattle.

Adaptation to a rigorous environment has also been achieved by the Criollo cattle. Tolerance of heat and of tick-borne diseases and an ability to subsist on poor forage enabled them to spread throughout Latin America and the southern parts of the United States. The Texas Longhorn and all the other Criollo races doubtless owe much of their form to natural selection. They provided excellent raw material for the great beef ranches of the southern and western United States. Criollo herds were upgraded by the use of imported Shorthorn and Hereford bulls, to such effect that today only a few herds of Texas Longhorns remain. In South America, zebu bulls were extensively crossed with Criollo cows and today only about 6,000 purebred Criollo cows remain.

In contrast, many other breeds owe their form to selection by man. In many of the White Park herds, colored calves were selectively killed, and at Chillingham today, all calves have red-brown ears, the result of the removal, during the 18th century, of all black-eared calves. The English Longhorn earned its place in history by being the subject of the first modern cattle breeding program. In about 1760 Robert Blakewell started to use the technique of intensive inbreeding to develop the large, rangy Longhorn into an early maturing beef animal. Unfortunately the new breed had a low milk yield and poor reproductive success and soon fell from favor.

A specialized form of selection is practiced on the De Lidia bulls of Spain and Portugal. Bulls and heifers are tested when a year old; to pass the test, the animal must repeatedly charge a horseman who pricks it with a lance. Records are kept of the results of these tests and there is much social prestige in being a successful breeder of these fighting bulls. From about 1860, De Lidia blood was introduced to the Camargue cattle to make them more aggressive (see boxed feature).

These breeds should not be seen as strange survivors from a bygone age. They are valuable reservoirs of genetic material as yet hardly tapped. The N'Dama, when crossed with Red Poll bulls, gave rise to the Senepol breed, one of the best humpless tropical beef breeds. The British White is being evaluated in Iowa as a crossing sire for the production of fattening steers from Angus cows. The carcass qualities of the English Longhorn are becoming widely appreciated. High priority is attached by the United Nations Food and Agricultural Organization to the conservation of the purebred Criollo. When crossed with zebu bulls they produce good beef animals but poor milking cows. Some purebred Criollo races are good dairy cattle and, when crossed with Jerseys, combine high productivity with tropical adaptability.   SJGH

Dairy and Mixed-use Cattle

There are now about 221 million dairy cows in the world producing about 480 million tons of milk each year. The major milk breeds are the closely related Friesians and Holsteins, Dutch and German in origin.

Friesians spread from Holland to all parts of Europe, replacing other indigenous dairy breeds, such as the Ayrshire and Red Danish from the 19th century onwards. They were taken to North America by early Dutch Settlers in the 17th century and a North American strain, known as the Holstein, followed a separate pattern of selective breeding in the mid-19th century. They bred

▶ **Serried Herefords.** OVERLEAF Beef cattle on the move.

◀ **In summer grazing,** a Brown Swiss ruminates beneath the peaks of the Alps. Characteristic of this high-altitude, mixed-use breed are short, black-tipped horns, a dished face and grayish-brown coat with a light ring around the muzzle. Lighter areas also appear on the ears and the lower legs.

◀ **Colombian corrida.** BELOW An ancient Spanish tradition thrives in the New World.

▼ **From time immemorial,** cattle of the Highland breed have been kept in the Western Isles and western Highlands of Scotland. Hardy animals with long, shaggy coats, they require no man-made shelter, even in the most severe winters. Cows calve regularly up to the age of 16 and over.

a dairy animal which is larger and higher yielding, but which has a poorer carcass conformation than Friesians for beef use. Under intensive farming conditions the breed can produce very high yields and an American Holstein holds the world record with an annual total of 55,661lb (25,247kg). In Israel, kibbutz herds kept in roofed paddocks and milked three times each day average over 17,600lb (8,000kg) each year.

Although being superseded by the Friesian in its native Britain, the Ayrshire has been exported worldwide and is now the predominant dairy breed in Scandinavia and in Kenya. Its adaptability, disease resistance and longevity have made it valuable in many crossbreeding programs, especially in developing countries.

Special recognition is given to the high-quality, creamy milk produced by the Channel Islands breeds. Developing independently on two small islands, the Jersey and Guernsey breeds are descended from cattle imported from northwest France. Both breeds are able to thrive in extremes of climate. A Guernsey cow accompanied Admiral Byrd's polar expedition in the 1930s and the breed is also found in the hot environments of Kenya and Egypt. Similarly the Jersey is farmed in such diverse regions as Alaska and the West Indies. Together with the Sahiwal, a zebu breed, it has been used to develop a new breed, the Jamaica Hope, able to produce good quantities of rich milk under tropical conditions.

Many important dairy breeds are mixed-use breeds, also useful for meat production. Typical of such breeding is the Normandy, descended from Viking cattle carried into France in the 9th and 10th centuries. Although its yield is relatively low, the milk is of consistently high quality and is used to make such famous cheeses as camembert and neufchatel. In addition it is a fast growing meat producer and even unwanted cows have well-fleshed carcasses.

Another dual-purpose breed, the Brown Swiss or Braunvieh, is specially adapted to life at high altitude. In its native Switzerland this hardy breed grazes alpine pastures up to 9,200ft (2,800m) above sea level throughout the summer, protected from mountain radiation by its pigmented skin and coat coloration. It has now spread to other mountainous countries such as Peru, where it is found in regions as high as 14,800ft (4,500m). A higher yielding lowland strain, averaging milk yields of 11,900lb (5,400kg) annually, is also widespread in North America.

The Simmental, known as the Fleckvieh in Germany or Pie-rouge in France, also combines an acceptable milk yield with very good meat production. It is one of the most numerous European breeds, made up of several different strains. One strain, the Montbeliarde, is a major French dairy breed averaging 11,900lb (5,400kg) annually at 3.7 percent butterfat. However, the emphasis is increasingly on meat production as other specialist dairy breeds develop. In tropical regions it is often crossed with native zebu cattle to give a hardy, fast growing animal such as the Simbrah developed in America. The Simental is now also widely used as a sire on specialist dairy breeds to give crossbred offspring with good beefing qualities.   SAE

Beef Cattle

Beef has always been much in demand, but even today, it remains something of a luxury. Some meat is supplied by dairy animals: calves, young bulls and culled cows. However, this production is insufficient, and many cattle populations have been selected specifically to produce meat. Beef breeds have been chosen exclusively for meat production. Others, whose milk-producing potential has been maintained, are called mixed-use breeds.

The Shorthorn was an early leader of popularity in beef breeds. Born in the Tees River Valley of northern England in the early 1600s, it has since improved cattle herds in more than 70 countries throughout the world. Widely adaptable "red-roan-'n-white" cattle arrived in the United States in 1783 landing in the State of Virginia under the label "Durham." Breed enthusiasts take special pride in listing present-day breeds which owe a share of their origin to Shorthorns-Santa Gertrudis, Beefmaster and Maine-Anjou forebears.

Beginning in Herefordshire, England, the Hereford is the most widespread beef breed. It has become symbolic of the beef industry in the United States due to the fact that a large part of American beef production carries a "whiteface" trademark. Herefords were bred early to Texas Longhorns. The Hereford has good mothering abilities and tender meat.

Attempts to produce the most meat possible have sometimes resulted in animals of extreme form. The congenital abnormality known as double muscling can be born into some French (Charolais), Belgian (Blue Belgian—a mixed-use breed) or Italian breeds (Piedmontese).

At first, animals were selected to be quickly and easily fattened while still

giving good-quality meat. This was particularly true for Herefords and Angus. When animals from draft breeds were used, they had to have large carcasses and muscular development, such as the Romagnola.

Some breeds can be traced back to selection for ritual sacrifice. This is why Chianina bulls look astonishingly like those in ancient Roman bas-reliefs. Selection of animals with great muscular development is not limited to humpless cattle. Especially in the tropical zones of the world, there is much use for meat production of zebu or of crosses between zebu and humpless cattle.

The qualities required of these animals, aside from their growth capacity, include hardiness, tolerance to under-feeding, ability to use rough forage, good reproductive potential and ease of handling. On the other hand, milk production is mediocre and varies from 1,700–3,500lb (800–1,600kg) in one eight-month period of lactation.

Animals are at pasture for at least the summer season, in a herd made up of a bull and cows with their calves. Artificial insemination is not very often used. The calves are separated from their mothers at 4–9 months of age. The great majority of males are castrated early in English-speaking countries, but late on the European continent, where entire males are fattened, making best use of their growth potential.

Outside Europe, beef breeding has been considerably developed, particularly in North America and Australia. Over the last few decades the consumer has been demanding leaner meat, and nutrition and management have been considerably improved, making it possible to better use animal growth potential. All this has led to the more and more frequent use of pure breeds of continental crossbreeds with great muscle development and without excessive fat. These same breeds are used more and more frequently for crossbreeding with dairy animals, particularly in eastern European countries.   PLN

Old and Local Cattle

The cow of an earlier age had to satisfy farmers with both milk and meat production, and she would also often find herself hitched up to the plow or wagon. For a long time attempts to increase production through breeding were rare. Often the best meat animals were sold for slaughter young, because a good price could be had, and they were lost for breeding. Animal husbandry was usually only one of several enterprises of the farmer and he could only give it a part of his attention.

In the early 19th century a systematic improvement of production began. Specialized milk or meat breeds were frequently imported from other areas. The new varieties required particular attention in terms of housing, feeding and care, and so other breeds were given up by many farmers. Old and local breeds had to make way for more productive ones, sometimes even if the difference was only minor.

Of the hundreds of breeds of cattle in the world, only about a dozen became internationally important as milk animals and another dozen or so as beef cattle. Others spread very little and only into neighboring areas or remained confined to localities where they were developed, decreased in numbers or became extinct. In Bavaria around 1860 there were 28 predominant breeds of cattle. In 1900 there were still ten and today there are only three.

Cattle were given up sometimes because it seemed more sensible to keep other species of animals—sheep instead of cattle, for example—on less fertile land. Also crop production became a specialization in some cases. In countries with flourishing economies, for example, increasing prosperity led to increased wine production in suitable regions, and the cattle of these regions were given up.

While it is true that some animals which would have been considered as a breed in the middle of the previous century would today only be classed as a subdivision of one

◀▲▼ **Beef and mixed-use cattle breeds.**
(1) Hereford bull (England: beef). (2) Chianina bull (Italy: beef). (3) Angus cow (Scotland: beef). (4) Charolais bull, showing heavily muscled hindquarters (France: beef). (5) Simmental cow (Switzerland: beef and dairy). (6) Blue Belgian cow (Belgium: beef and dairy). (7) Limousin bull scratching neck (France: beef).

breed, the tendency towards a decrease in breed numbers—and thus towards a decrease in genetic variation—is apparent. Even among old and local breeds only the most productive animals remained and these changed in time. Breeds have not been static. They have changed their appearance, size and weight, and above all have increased their productivity. This occurred in part, although not always, with the crossing of different breeds.

Intensive breeding for production alone, with its accompanying standardization, can result in a loss of genes. It then becomes difficult to adapt to changing production conditions through breeding. The genetic resources of old and local breeds may be needed yet. They require less high-energy feed and may need to be crossed into the high-production breeds in a future world short of cereal grains. Even today they have advantages for developing countries with less intensive conditions of production. Old and local breeds can offer longevity, minimal calving problems, good mothering abilities, greater disease resistance and resistance to rigorous climate.   HHS

Zebu Cattle

Zebu cattle differ from humpless cattle in having a longer, narrower skull, a hump and long, drooping ears. They usually have a heavy dewlap and pronounced penis sheath in the male or an umbilical skinfold in the female. Not all humped cattle are zebus, because crosses between them and humpless cattle, called sanga, can also carry a hump (see 47–49).

The hump, better developed in bulls than in steers and cows, is already discernible in a 50-day fetus. It is composed of muscle, or muscle and fat mounted above the withers, and is quite unlike the humps of the gaur, mithun or banteng, which are supported by spiny protrusions from the vertebrae.

Zebu cattle, which can withstand high temperatures, ticks and many diseases of the tropics, accompanied their owners on migrations in early times into Africa, Madagascar, the Philippines, Indonesia, Southeast Asia, China and the Near East. In more recent times they have been used in the development of North and South American cattle; but their homes and centers of diversity are in India and Pakistan.

Zebu cattle are of huge importance in India, where cattle number 200 million. More power is provided by the bullock population than by all the Indian electricity generating stations. Cattle and buffalos provide two-thirds of the energy used on Indian farms; human muscle provides 15 percent

▼► **Old and local cattle breeds.** (**1**) Dexter bull (Ireland: dairy). (**2**) Aurochs bull (wild type *Bos primigenius*). (**3**) Mongolian cow (China: draft and dairy). (**4**) Romagnola cow (Italy: draft and beef). (**5**) Finn cow grooming with tongue (Finland: dairy). (**6**) Telemark cow (Norway: dairy). (**7**) Belted Galloway cow (Scotland: beef). (**8**) Groningen bull ruminating (Netherlands: beef and dairy). (**9**) Eringer cow (Switzerland: dairy and cow fighting).

Fighting Cattle of Switzerland

On hearing the term multiple-purpose cattle, one thinks of animals producing milk, meat and possibly draft power at the same time. A major reason for the maintenance of the Eringer breed, which is found exclusively in the canton of Wallis in Switzerland, however, is fighting ability. This fighting is not bullfighting against people, but fighting against one another, and involves only the females. The animals are divided into five categories by age and weight. About 12 are put into the arena simultaneously. The best cows from the seven regional championships qualify for the final round of fighting. The strongest of each category is finally determined by several qualifying rounds. Often this takes a long time, as the animals are very aggressive and fight bitterly and stubbornly. The victors of the first four categories determine the "queen" among themselves. A "queen" not only brings her owner much prestige, but her value increases many-fold compared to the other cows.

HHS

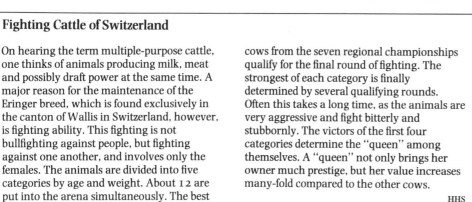

and electric power and oil provide the rest.

In India, cattle are far more important as milk producers than as beef animals. In attempts to increase milk production local cattle are often crossed with imported Jerseys or Friesians, which are high yielding but not well adapted to local conditions. This is causing degeneration of the stock. It is now a policy to create pure herds of the zebu breeds and gather data on their performance so the best use can be made of them. The vast majority of Indian cattle are nondescript, owing their ancestries to mixtures of the 26 known zebu breeds.

Of these breeds, Kankrej cattle have always been noted as draft animals. On a good road, a pair of bullocks can move a 4,000lb (1,800kg) load 25mi (40km) in a working day of 10 hours, and they are good for all types of field work. Between 1875 and 1921 they were exported to Brazil where they gave rise to the Guzerat beef breed. Kherigarh cows are poor milkers but the bullocks are noted for their endurance. In contrast, the Red Sindhi is not widely known as a draft breed, but is probably the best Indian milk breed. It has been used to upgrade local breeds in Malaysia, Burma, Sri Lanka, the Philippines, Japan and Brazil. The Haryana is one of the most important dual purpose (milk and draft) breeds in northern India; a pair of bullocks can be expected to provide all the draft effort needed to run a holding of 14 acres (5.5ha). When used extensively on hard roads they are shod.

Like the Kankrej, the Gir has found favor in Brazil where it is called the Gyr. In India it is reputed to be the best zebu beef breed but few are ever slaughtered. Gyr and Guzerat cattle, with the Nellore (see over)

together formed the Indo-Brazilian breed.

The Tharparkar is remarkable for its hardiness and heat resistance, being a good milk and draft breed in a rigorous semi-desert native environment. Ongole cattle were developed in a more benign environment and adapted well to Brazilian conditions where they formed the Nellore breed, widely used in Latin America.

The Hallikar, like the Haryana, is very important in India but has made no impression elsewhere. It is one of the best all-round draft types in southern India, able to cover 30–40mi (48–64km) a day on rough roads; but it has little milking capacity. The Lohani is also of great importance locally as a working or pack animal and has potential for development as a milk and draft breed for areas where large-bodied cattle are not needed.

The Boran is an African zebu. It is the outstanding beef breed of East Africa and is tolerant of heat and of many diseases but not of trypanosomiasis. It takes a long time to mature, but since 1920 selective breeding has led to earlier maturity and a good beef conformation. Zebu cattle are well adapted

to the drought conditions which often prevail in East Africa; they stop eating when deprived of water and use their fat. This reduces the urinary and fecal water loss. Zebu feces become much drier when water stress occurs—they contain 6.5oz (190g) water per 3.5oz (100g) dry matter. In contrast feces of Hereford steers kept under the same conditions contain 10.5oz (300g) water per 3.5oz (100g) dry matter. Herefords need twice as much water as zebus.

Early on, these qualities became known to American cattle raisers. The Brahman is the standard American zebu breed and was created from fewer than 300 individuals of the Guzerat, Gir and Nellore breeds. This process began in 1854 when a Louisiana farmer was given two Indian bulls. More were imported directly from India, some were bought from circuses and others brought from Brazil, via Mexico. These bulls were used on the big herds of Texas Longhorn descent in the southern United States which by this time had been upgraded to Herefords and Shorthorns. This cross has been very successful because of the heat tolerance and disease resistance conferred

▲ **Admirably suited to dry and tropical conditions,** these zebu Nellore cattle have been bred for beef production in Brazil, displacing Criollo cattle descended from southern European breeds. The hump over the withers, the long face, drooping ears and heavy dewlap are characteristic of the zebu breed type.

▶ **Sacred Watusi cattle.** ABOVE The Inyambo strain, revered by the Watusi herdsmen who keep it, is a variety of Ankole with horns that may measure 7.5ft (2.3m) between the points. As the traditional breeders of Ankole cattle begin to see the value of their animals in terms of milk and meat, they are replacing them with disease-resistant, higher-producing zebu.

▶ **Initiation ceremony.** BELOW A young Dinka in the southern Sudan sings to the ox which he has been given on entering manhood. Its distinctive color and shape determine his own adult name. Like other sanga cattle, this animal carries its hump farther forward than zebus do.

▶ **Rice-field workers on Bali,** OVERLEAF the only location where domesticated banteng are the predominant form of cattle. Bali cattle are more important as draft animals, but also supply beef to the Hong Kong market.

by the Brahman and the hybrid vigor resulting from the crossing of two genetically distant strains, but the beef conformation is not yet ideal. Brahman bulls are being used in more than 60 countries in the tropics and subtropics.   SJGH

Sanga and other Cattle

Sanga cattle originated when zebu cattle (see above), which had entered Africa from Aden probably about 200BC, were crossed with the humpless cattle which had already been introduced to Africa. Typically, Sanga cattle have long horns and a hump located in front of the withers. (The zebu hump is on top of the withers.) As in the zebu it is more marked in bulls. The most notable sanga are the White and Red Fulani, the Africander, the Ankole and the Nguni.

Chinese yellow cattle also arose from a zebu-humpless cross but it is not known when or where. In contrast, the Santa Gertrudis and the other modern crosses of Zebu breeds with humpless stock are well documented. Nothing is known of the history of domestication of the gaur (*Bos gaurus*) and the banteng (*Bos javanicus*), which gave rise to the mithun and the Bali cattle.

The White Fulani is the outstanding breed of northern Nigeria and has potential for development as a milk–meat breed. The cattle-raising Fulani are intensely proud of their animals and their whole culture revolves around them. The Red Fulani is Nigeria's main beef breed.

Ankole cattle, too, are intimately associated with the culture of their owners, the Watusi. Their main value is as status symbols. The extraordinary horns—40in (100cm) or more in length in the Red Fulani, 60in (150cm) in the Ankole with a diameter of 6in (15cm) at the base and a gap of 80in (200cm) between the tips—must be the product of human selection. In contrast, the **Nguni** cattle have much more modest horns. In Matabeleland, southern Africa, selection was concentrated on a particular side color pattern and this was the foundation for an excellent beef breed which is also noted for milk production.

The Africander is the most important commercial breed in South Africa. It was developed by the Dutch settlers from the sanga cattle kept by the Hottentots. Exactly like the other African sanga breeds it is part of the heritage and patrimony of the people who keep it. Purebred Africanders must have the sloping rump of oxen bred for hauling wagons across the veldt, rather than the rounded rump which, admittedly, would make it a better beef animal.

The Chinese yellow cattle are a breed group or type rather than a single breed. A principal use of this group is for draft, and when a bull is crossed with a female yak a valuable draft, milk and pack hybrid results, the Pien Niu; it has the hairy coat and varied color of the yak. Male hybrids are sterile, females fertile.

Modern cattlemen have created their own sanga breeds, by crossing zebu bulls with humpless cows. The Santa Gertrudis was developed on the King Ranch, Texas, in a successful attempt to combine the hot-weather adaptations of the Brahman (see pp46–47) with the beef conformation of the Shorthorn breed (see p40). By 1940 purebred Santa Gertrudis bulls were being produced in large numbers and used to upgrade the Shorthorn herds, and now they are to be found in hot countries all over the world. Other breeds have been developed in the same way. The Brangus (developed in Louisiana and Oklahoma) is 37.5 percent Brahman and 62.5 percent Angus, and the Braford (Texas and Louisiana) is 37.5 percent Brahman and 62.5 percent Hereford. In Australia, the Shorthorn, which originally came from Britain, has done very well in the tropical north and has become adapted to heat stress and tick-borne diseases; attempts are being made to improve it further, crossing in Brahman bulls to produce the Droughtmaster.

The mithun and Bali cattle have completely different lineages. One must go back from 1–3.5 million years to find the ancestor they have in common with cattle. In spite of their long evolutionary history apart, both will hybridize with cattle; male offspring are infertile, females fertile. Mithun live free in the woods and some return to the villages at night; others are controlled and kept near at hand by the provision of salt which they crave. They have considerable potential as meat or draft animals; at present there are 150,000 in India, Bhutan, Bangladesh and Burma. They are unique among domesticated mammals in that they are used only for sacrifice on special religious and social occasions.

Bali cattle are, apart from elephants, the only mammal to have been domesticated in the humid tropics. They are very much like their wild ancestor, the banteng, which is to be found in scattered locations in Southeast Asia. There are only about a million Bali cattle. They have potential for development as a draft and beef breed for the humid tropics where they may prove to be better than the zebu cattle, which excel in the drier areas.   SJGH

CATTLE BREEDS

Long-horned Breeds

White Park and Chillingham

Originated: England, Scotland and Wales. Only found in Britain principally in these herds: Chillingham (N England)—purebred; Vaynol (N Wales), Cadzow (Scotland), Chartley (England), Dynevor (S Wales). Coat: white; Chillingham cattle have red-brown ears and red-brown or black markings on face, neck and withers; the others have black ears and black spots and patches. Horns: variable in shape; medium long. Features: compact, well muscled, rather short-legged. wt: cow 617lb, bull 661lb. ht: cow 43in, bull 45in (Chillingham). Cow 1,322lb, bull 1,432lb (other herds). Ornamental livestock; formerly hunted.

British White

Small herds in Britain, Australia, USA (where sometimes known as White Park). Originated: England, developed from White Park with Swedish Mountain blood added. Coat: white, ears black, small black patches and spots on many animals. Horns: none (polled breed). Features: bulky, well muscled. wt: cow 1,322lb, bull 1,432lb ht: cow 51in, bull 59in. Beef and dairy breed.

Highland

Highland, Kyloe or West Highland

Originated: NW Scotland. Farmed: Scotland, USA (Dakotas and Alaska), Canada. Coat: usually red-brown but also brindle, yellow-tan, black; long hair covers eyes, tail and extends down legs. Horns: large, curving forward and out (bull); upward (cow). Features: short legged, compact. wt: cow 881lb, bull 992lb. ht: cow 47in, bull 55in. Beef breed.

De Lidia

Farmed and originated: Spain and Portugal. Coat: black or dark gray, occasionally brindle or mixed. Horns: spread up either in a single curve or with slight inward curve at top. Features: deep-chested, barrel-shaped body; strongly muscled. wt: cow 661lb, bull 771lb. ht: cow 43in, bull 47in. Rodeo and fighting bull.

Camargue

Farmed and originated: Rhone Delta of SE France; De Lidia blood added. Resembles De Lidia. wt: cow 551lb, bull 661lb. ht: cow 43in, bull 45in. Rodeo and fighting bull.

Criollo

Local breeds throughout Latin America and S USA. Originally imported from Spain to Hispaniola end of 15th century. Coat: varies from solid tan (like Jersey) to solid cherry-red; also black, white and all mixtures (few brindles). Horns: medium, unswept. Features: very variable, but generally has barrel-shaped body and long legs. wt: cow 485–1102lb, bull up to 1,212lb. ht: cow 43–59in, bull 47–59in. Mostly dairy; some draft or beef.

Texas Longhorn

Small herds in Oklahoma and other western states of the USA. Originated: Mexico, Texas, New Mexico from Criollo stock. Coat: very variable; speckled, patched black on white or red on white, brindle, occasionally solid red, tan, white or black. Horns: widespread usually with some upturn usual span up to 47in (cow), and 67in (bull). Features: rangy, long-legged, with long face and narrow head wt: cow 793–992lb, bull up to 1,587lb. ht: cow 47in, bull 51in. Beef breed.

N'Dama

Farmed and originated: coastal rain forest W of Nigeria. Coat: usually tan (like Jersey). Horns: spread up and out, sometimes lyre-shaped; some varieties have strong polled tendency. Features: compact, well muscled with fine bone structure. wt: cow 595lb, bull 705lb. ht: cow 43in, bull 47in. Beef and draft breed.

English Longhorn

Farmed and originated: NW and central England and Ireland. Coat usually dark red brindle; white line along back and down tail. Horns: long, spreading outwards and upwards, sometimes forwards and inwards. Features: bulky, rather short legged. wt: cow 1,432lb, bull 1,543lb. ht: cow 47in, bull 51in. Beef breed.

Dairy and Mixed-use Breeds

Ayrshire

Europe, N and C America, Australasia, Far East, S Africa. Originated SW Scotland. Coat: red or brown with varying amounts of white. Features: large upward-curving horns. wt: cow 1,157lb, bull 1,962lb. ht: cow 52in, bull 55in. Milk yield p.a. 11,020lb at 3.9% bf.

Brown Swiss

Brown Swiss, Brown Alpine or Braunvieh

Europe, N, C and S America, Mediterranean, Asia, S Pacific, Africa. Originated Switzerland. Coat: grayish-brown, lighter around muzzle and on lower legs. Features: short, forward-curving horns; face slightly dented between eyes. wt: cow 1,334lb, bull 2,226lb. ht: cow 54in, bull 58in. Milk yield p.a. 9,040lb at 4% bf.

▲ ▶ ▼ **Dairy and longhorn breeds.** (1) Criollo bull (South America, southern USA: dairy, draft and beef). (2) N'Dama cow (West Africa: beef and draft). (3) English Longhorn cow (beef). (4) Texas Longhorn cow (beef). (5) Camargue bull pawing ground in threat (France: bullfighting). (6) Meuse-Rhine-Ijssel bull (Netherlands: dairy). (7) De Lidia bull threatening with horns (Spain and Portugal: bullfighting). (8) Brown Swiss cow (Switzerland: dairy). (9) Jersey cow sleeping (Channel Islands: dairy). (10) Normandy cow ruminating (France: dairy). (11) Red Danish cow lowing (Denmark: dairy). (12) Friesian cow licking (Netherlands: dairy). (13) Dairy Shorthorn cow (England: dairy).

Dairy Shorthorn

Europe, N America, Australasia. Originated N England. Coat: red to white with many roans. Features: short, forward-curving horns. wt: cow 1,257lb, bull 2,094lb. ht: cow 53in, bull 56in. Milk yield p.a. 10,360lb at 3.6% BF.

Friesian/Holstein

Europe, N and S America, Asia Minor, Japan, Australasia, S Africa, Israel. Originated N Holland, N Germany. Coat: black and white or red and white, white tail switch and lower legs. Features: short, outward-pointing horns. wt: cow 1,433lb, bull 2,205lb (Friesian); cow 1,488lb, bull 2,403lb (Holstein). ht: cow 55in, bull 58in (Friesian); cow 56in, bull 63in (Holstein). Milk yield p.a. 12,345lb at 3.8% BF (Friesian), 13,888lb at 3.8% BF (Holstein).

Guernsey

Europe, N and C America, S Africa, Asia, Australasia. Originated Guernsey, Channel Islands. Coat: fawn with or without white markings, white tail switch. Features: long, narrow head, forward-curving horns wt: cow 1,100lb, bull 1,653lb. ht: cow 49in, bull 56in. Milk yield 9,038lb p.a. at 4.7% BF.

Jersey

Europe, N, C and S America, W Indies, Africa, Australasia. Originated Jersey, Channel Islands. Coat: fawn to brownish black with or without white patches. Features: short head, dished (concave) face, forward-curving horns. wt: cow 860lb, bull 1,500lb. ht: cow 47in, bull 49in. Milk yield p.a. 8,598lb at 5.2% BF.

Meuse-Rhine-Ijssel

Europe, N America. Originated S Holland. Coat: red and white. Features: short, forward-curving horns. wt: cow 1,455lb, bull 2,248lb. ht: cow 53in, bull 57in. Milk yield p.a. 11,465lb at 3.7% BF.

Normandy

Europe. Originated NW France. Coat: white with dark red or brown streaks or patches, dark rings round the eyes. Features: short, outward-pointing horns. wt: cow 1,432lb, bull 2,205lb. ht: cow 55in, bull 59in. Milk yield p.a. 8,377lb at 4.2% BF.

Red Danish/Red Polish

Europe, N America. Originated Denmark. Coat: deep red. Features: horns curve forward and slightly downward. wt: cow 1,432lb, bull 2,205lb. ht: cow 52in, bull 58in. Milk yield p.a. 9,920lb at 3.7% BF.

Simmental

Farmed: Europe (almost all mountainous zones) N and S America, Africa, Australia. Originated: N Switzerland. Coat: dun-red or leather-yellow and white. Horns: curving forward and outward. Features: very good body conformation; good milk production (8,818–11,020lb p.a. at 4.0% BF). Used to improve performances of other breeds, particularly in South Africa. wt: cow 1,653lb, bull 2,380lb. ht: cow 54in, bull 57in.

White-blue Belgian

Farmed and originated: Belgium. Coat: white, blue-gray or blue. Features: very well shaped; high proportion of double-muscled animals. Used for cross-breeding. Cows produce about 8,818lb of milk p.a. wt: cow 1,653lb, bull 2,535lb. ht: cow 54in, bull 59in.

Maine Anjou

Farmed: worldwide. Originated: western France, by crossing of local breeds with Shorthorn bulls. Coat: red, red and white, or roan. Features: includes the world's biggest bull (3,880lb). Used frequently for crossbreeding. Milk production limited to about 6,614lb at 3.6% BF p.a. wt: cow 1,984lb, bull 2,756–3,880lb. ht: cow 55in, bull 57in.

Beef Breeds

Hereford

Most widespread beef cattle in the world. In 1970 horned and polled varieties accounted for 45% of the meat breeds found in the USA. Originated: Herefordshire, England. Coat: very distinctive: red with white head and underside. Features: selected for meat production since 18th century; extremely hardy, used in the cold areas of Canada and even in the arid regions of Australia; fattens rapidly and yields very good quality meat. wt: cow 1,190lb, bull 1,840lb. ht: cow 51in, bull 53in.

Angus

Second most numerous beef breed in world. Originated: NE Scotland. Coat: black. Features: polled; small-sized but with very good conformation; growth performance still limited compared to other breeds, but yields good quality meat. wt: cow 1,146lb, bull 1,764lb.

Shorthorn

Once very important, particularly in Argentina and USA, now considerably reduced in numbers. Originated: NE England. Coat: red, roan or white, sometimes red and white or roan and white. One of the first, after the Longhorn, to be selected according to modern principles. wt: cow 1,212lb, bull 1,764lb.

Murray Gray

Farmed: Australia, USA, Britain, N Zealand, S Africa. Originated: Victoria, Australia, from a light roan Shorthorn cow crossed with an Angus bull in 1905; dominant gray color retained in some later crosses with Angus. wt: cow 1,102–1,642lb, bull 1,697–2,000lb.

Charolais

Farmed: worldwide. Originated: EC France. Coat: white. Features: efficient feeder with high growth rates; limited cross-breeding with white Shorthorn; females with high milk-producing potential; very successful both as a purebread and for crossbreeding. wt: cow 1,653lb, bull 2,402lb. ht: cow 54in, bull 56in.

Limousin

Farmed: France and N America. Originated: Limousin region of France. Coat: dark yellowish-brown. Features: efficient feeder with high growth rates. Male calves used for veal or slaughtered at 880lb following short fattening period. Females have world reputation for reproductive qualities and easy calving in spite of their very good conformation. wt: cow 1,322lb, bull 2,094lb. ht: cow 53in, bull 57in.

Romagna

Farmed and originated: Italy. Coat: gray with dark patches in front and on legs. Features: large size, good conformation. Long used for draft, today for meat production in many countries. wt: cow 1,410lb, bull 2,425lb. ht: cow 59in, bull 62in.

Chianina

Farmed: Italy and N America. Originated: central Italy. Coat: white with black spots and mucosa, black eyelashes; calves reddish yellow. Horns: in bull rather thick, extending outward form the sides of the head with a slight curve; in cow usually turned upward. Features: large size, good conformation; largest and most impressive bulls in the world; meat of excellent texture. Frequently used in crossbreeding. wt: cow 1,873lb, bull 2,645–3,968lb. ht: cow 62in, bull 71in.

Piemontese

Farmed: Italy. Supposed to have originated by crossing zebu with northern Italian breeds. Coat: short, smooth light gray on the females, dark gray on the males. Eyelashes, border of the ears, tail switch, as well as hooves and horn tips black. Bull shows dark shadings around the eyes, nape of the neck, forequarters, and belly. Calves reddish yellow first months of life. Features: horned; very good muscling, frequently double layered wt: cow 1,102–1,323lb, bull 1,763–2,204lb. ht: cow 51in, bull 55in. Milk yield p.a. 5,290lb at 3.7% BF.

Romagnola

Used internationally in cross-breeding programs with milk and beef breeds. Originated: N Italy. Coat: cow white; bull light to dark gray, usually darker in the fore parts of the body, muzzle black, tail switch black; calves light brown. Horned. wt: cow 1,763lb, bull 2,645–3,086lb. ht: cow 62in, bull 67in.

Old and Local Breeds

Angler

Farmed: Germany. Originated: Angeln Peninsula in N Germany, crossed with the Red cattle of Poland, the Soviet Union and Syria. Coat: rather dark red. Horns: short and forward curving on female. wt: cow 1,212–1,322lb, bull 1,874–2,204lb. ht: cow 50–52in, bull 54–59in. Milk yield p.a. 11,023lb at 4.6% BF.

Hinterwälder

Farmed and originated: S Germany. Coat: yellow-white or reddish-white with white head. Features: horned; smallest breed of cattle of middle Europe; well adapted to living on barren, mineral-poor mountain meadows; good longevity; very spirited. wt: cow 770—880lb, bull 1323–1,653lb. ht: cow 45–47in, bull 49—51in. Milk yield p.a. 5,290lb at 4.2% BF.

German Yellow

Farmed: Germany, N and S America, S Africa, United Kingdom. Originated: S Germany by the crossing of Simmental and Shorthorn blood into the native breed. Coat: uniform yellow with light muzzle. Features: horned; powerful bones, heavy muscling in a breed originally bred for good work capacity. The draft oxen were prized even in distant regions. wt: cow 1,433–1,763lb, bull 2,535–2,865lb. ht: cow 53–55in, bull 59–61in.

Aurochs (back bred)

Several herds in zoos and on farms in Germany. Originated about 50 years ago in Germany by crossing Hungarian Steppe cattle, Scottish Highland, German Braunvieh, Angler and others. Coat: bulls black with yellowish midline stripe on back; cows reddish-brown with darker neck; mouth area white in both sexes. Horns: long, thick, swung forward with black tips. wt: cow 1,212lb, bull 1,653lb. ht: cow 51in, bull 55in.

Pinzgauer

Farmed: Austria, S Germany, Czechoslovakia, Yugoslavia, S Africa. Originated: Pinzgau Valley in Austria, 17th century, by the crossing of Simmentals with a solid red regional type. Coat: mostly dark red, chestnut brown; white stripe begins at the withers, runs along the back, down the back of the rear legs and along the belly to the brisket. White bands on the forelegs and lower thighs. Features: white tail; horned. wt: cow 1,323–1,543lb, bull 2,204–2,535kg. ht: cow 49–53in, bull 54–60in.

Eringer

Farmed and originated: Canton of Wallis, Switzerland. Coat: solid, dull black, sometimes shading to a reddish tint over the hips and sides; thin line of red hair down spine quite common. Horns: on cow medium-sized, upturned and often tilted backward toward the tips; on bull out-thrust, nearly straight, short and thick. Head: short, broad with concave bridge of nose. Features: requires little care; belligerent mountain breed with fine boning and good muscling. wt: cow 880–1,212lb, bull 1,323–1,653lb. ht: cow 47–51in, bull 53in. Milk yield p.a. 7,054lb at 3.8% BF.

Groninger

Farmed and originated: Netherlands. Coat: black or red with small white undermarkings and white head; ring or patch around eyes, same color as coat; pasterns and switch may be white. Horns: in bull small and out-thrust with a forward curve; in cow with sharp forward curve. wt: cow 1,323–1,543lb, bull 2,204–2,535lb. ht: cow 51–54in, bull 56in. Milk yield p.a. 11,463lb at 4% BF.

Lakenvelder
Lakenvelder, Veldlarker or Dutch Belted

Farmed: Netherlands and USA. Originated: Netherlands. Coat: black or red with clearly defined white belt. Horns: long relative to diameter. Features: hair fine and soft; striking appearance. wt: cow 880–1,410lb, bull 1,598–1,984lb. Milk yield p.a. 11,464lb at 3.9% BF. Belted cattle also existed in 19th century Bohemia (now Czechoslovakia). The Sheeted Somerset of England is extinct. The Belted Welsh in Wales and the Belted Galloway in Scotland survive in small numbers.

Kerry

Farmed: Ireland, United Kingdom. Originated: County Kerry in Ireland. Coat: uniform black, occasionally with white spots in the udder area; quite rough in the winter. Horns: rather large for size of animal. Features: exceptionally long lived. wt: cow 880lb, bull 1,100lb. ht: cow 48in, bull 51in. Milk yield p.a. 7,054lb at 3.9% BF.

Dexter

Farmed: United Kingdom. Originated: Ireland. Coat: usually black, sometimes dark red or yellow; always solid, with only very minor white markings on the udder. Horns: rather small and thick, growing upward on the cow and outward with a forward curve on the male. Features: extremely short legs; conspicuous strong angle in the hind legs. Dwarf variation of Kerry, smallest breed in the British Isles. wt: cow 595–705lb, bull 992lb. ht: cow 43in, bull 47in. Milk yield p.a. 5,511lb at 4.5% BF.

Galloway

Farmed: United Kingdom, N America, New Zealand, Germany. Originated: Scotland. Coat: solid black, often with a brownish tinge; in winter quite rough but sheds to smooth, fine undercoat in the summer. Belted Galloways, a separate breed, have a white band around body. Features: polled; extremely hardy hill cattle, maintained in the open the year round; considered the oldest beef breed in Britain. wt: cow 992lb, bull 1,323lb.

Icelandic

Farmed and originated: Iceland. Coat: more variable than in any other cattle population in Europe; solid colored blacks, browns, reds and grays occur, as well as animals spotted with these colors and white. There are also almost completely white animals with

little spots of pigment as well as brown-black brindling. Features: wide variation in size and conformation; horned, hornless, and animals with scurs occur. wt: cow 880–1,102lb, bull 1,433–1,874lb. ht: cow 49–53in, bull 53–57in.

Telemark

Farmed and originated: Norway. Coat: reddish-brown divided by a white band along the spine and white bottom-line; face, tail and legs below the knees white. Horns: large, upswept, and curved in cow; pointing forward and shorter and thicker on bull. wt: cow 880–1,322lb, bull 1,543–1,764lb. ht: cow 43–47in, bull 53in. Milk yield p.a. 11,023lb.

Finncattle

Farmed and originated: Finland. Coat: light reddish-brown (West Finncattle) or tan to reddish-brown, usually with a wide, jagged white topline and bottomline with much white on the face and legs (East Finncattle). Mainly polled. wt: cow 880–1,102lb, bull 1,543–1,764lb. ht: cow 47in, bull 51in. Milk yield p.a. 11,023lb at 4.6% BF.

Kholmogor

Farmed: N Soviet Union. Originated: Russia. Coat: basic hair coloring black-pied (from white with a small number of spots to black); red-pied and brown-pied are also sometimes found. wt: cow 992–1,212lb, bull 1,873lb. ht: cow 52in, bull 57in. Milk yield p.a. 7,716lb at 3.7% BF.

Mongolian

Farmed: China. Originated: N and W China. Coat: almost solid black, brown or yellow, often with some white spotting. Horns: slender, thrust up and forward. wt: cow 595lb.

Tibetan

Farmed: China. Originated: Tibetan plateau. Coat: varies widely. Used as pack and draft animals. Milk yield p.a. 220lb for a 105-day lactation period.

Zebu Breeds

Kankrej

Farmed and originated: SE Rann of Kutch, Gujarat, India. Coat: gray or steel black; forequarters, hump and hindquarters darker, especially in male. Tail switch black. Horns: strong, lyre-shaped. Hump well-developed; dewlap thin, sheath and umbilical fold prominent. Features: one of heaviest Indian breeds; large, drooping, open ears are very

characteristic wt: cow 948lb, bull 1,323lb. ht: cow 51in, bull 63in. Particularly good draft animals; fair milkers.

Kherigarh

Farmed and originated: N Kheri, NC Uttar Pradesh, India. Coat: white or gray. Tail switch white. Horns: thin and upstanding, to 16in (bulls), shorter in cows. Hump well developed; dewlap thin, sheath and umbilical fold not prominent. Features: limbs light, body broad and deep. wt: cow 705lb, bull 1,058lb. ht: cow 51in, bull 51in. Active draft animals. Poor milkers.

Red Sindhi

Farmed and originated: Karachi, Hyderabad and SW Sind, Pakistan. Coat: red, ranging from dark red to dun. Horns: short, thick at base, emerge laterally and curve upwards. Hump medium; dewlap, sheath and umbilical fold moderately developed. Features: small, with deep, compact frame. wt: cow 705lb, bull 992lb. ht: cow 47in, bull 51in. Primarily milk breed.

Haryana

Farmed and originated: E Punjab, India. Coat: white or gray, fore- and hindquarters darker in bulls. Tail switch black. Horns: fine and short (6–8in in bull), emerge horizontally, turning upward. Hump large; dewlap deep, sheath and umbilical fold not pronounced. Features: well proportioned, compact; head held high, with bony prominence at center of poll. wt: cow 770lb, bull 1,102lb. ht: cow 51in, bull 55in. Useful for draft, fairly good milkers.

Gir

Farmed and originated: S Kathiawar, Gujarat, India. Coat: white with dark red or brown patches; sometimes entirely red. Horns: emerge at base of crown, curve down and back, then upward. Hump medium; dewlap and umbilical fold medium; sheath large. Features: very prominent, broad forehead; ears long and drooping with characteristic notch near tip. Unusual horn shape. wt: cow 838lb, bull 1,190lb. ht: cow 51in, bull 55in. Reputed best beef breed in India; fairly good milkers; heavy draft.

CONTINUED ▶

Tharparkar

Farmed and originated: SW Sind, Pakistan. Coat: white or gray; occasionally Red Sindhi and Gir influence seen. Horns: set well apart, curving gradually up and out, shorter and straighter in bull. Hump, dewlap and sheath medium; umbilical fold prominent. Features: medium size, compact. wt: cow 838lb, bull 1,190lb. ht: cow 51in, bull 51in. Has potential for milk production, exploiting hardiness and heat resistance.

Ongole

Farmed and originated: Guntur-Nellore, S Andhra Pradesh, India. Coat: white, dark gray on head, neck and hump (bull); red or red and white animals occasionally seen. Horns: short and stumpy, directed outward and back. Hump big; dewlap, sheath and umbilical fold prominent. Features: forehead broad between the eyes and slightly prominent. wt: cow 948lb, bull 1,256lb. ht: cow 51in, bull 59in.

Hallikar

Farmed and originated: Tumkur, Hassan and Mysore, India. Coat: gray to dark gray, dark shading on fore- and hindquarters. Tail fine with black switch. Horns: emerge close together from top of poll and are carried back each in a straight line for about half their length, then bend inward. Hump and dewlap medium; sheath and umbilical fold undeveloped. Features: characteristic horns and head; prominent, rather bulging forehead; ears pointed, held sideways. wt: cow 705lb, bull 992lb. ht: cow 47in, bull 55in. One of the best all-round draft types in S India but poor milkers.

Lohani

Farmed and originated: Loralai district of Baluchistan, Pakistan. Coat: red with white patches. Horns: short, slender, emerge sideways. Hump medium, dewlap long; sheath and umbilical fold undeveloped. Features: small, face slightly convex between eyes. wt: cow 440lb, bull 551lb. ht: cow 39in, bull 43in. Used as pack animals, fast draft; fair milkers.

Boran

Farmed and originated: N Kenya, S Ethiopa, SW Somalia. Coat: solid red or solid whitish-gray. Horns: small and thick, tendency to be polled. Hump large; dewlap, sheath and umbilical fold prominent. Features: well muscled. wt: cow 529lb, bull 660lb (Boran): wt: cow 992lb, bull 1,323lb. ht: cow 47in, bull 49in. (Improved Boran). Beef breeds.

Brahman

Farmed and originated: S USA. Coat: solid light to medium gray or solid medium-dark red; darker fore- and hindquarters (bull). Horns: short and thick, often upturned. Hump large; dewlap, sheath and umbilical fold prominent. Features: deep, broad body; ears long and drooping. wt: cow 1,323lb, bull 1,763lb. ht: cow 59in, bull 67in. Beef breed.

Sanga and other Cattle

Chinese Yellow

Farmed and originated: SE China. Coat: yellow-tan (like Jersey). Horns: characteristically short and thick. Hump small and poorly defined; dewlap quite prominent but sheath and umbilical fold not enhanced. Features: well-muscled forequarters, hindquarters narrow but heavily boned: wt: cow 770lb, bull 992lb. ht: cow 45in, bull 51in. Draft breed.

White Fulani

Farmed and originated: N Nigeria. Coat: white with black ears, muzzle and feet; commonly with black or red spots. Horns: medium length, thick, rising up and forward. Hump large; dewlap, umbilical fold and sheath quite prominent. Features: well proportioned, strongly built. wt: cow 661lb, bull 992lb. ht: cow 47in, bull 59in.

Red Fulani

Red Fulani or Red Bororo

Farmed and originated: N Nigeria. Coat: rich red-brown. Tail switch and muzzle often light colored. Horns: long (29–39in). Hump well developed; dewlap, sheath and umbilical fold also prominent. Features: rangy, long limbed. wt: cow 660lb, bull 880lb. ht: cow 47in, bull 55in. Beef breed.

Ankole

Ankole, Watusi or African Longhorn

Farmed and originated: E Africa, between Lake Mobutu Sese Seko and Lake Tanganyika (Uganda, Rwanda, Kenya, Burundi and Tanzania). Coat: usually dark red, also tan, black or pied. Horns: world's largest, sweeping up and out (39–49in). Hump small; dewlap, sheath and umbilical fold moderately developed. Features: long-limbed, rather lightly built. wt: cow 660lb, bull 880lb; ht: cow 47in, bull 55in. Kept by seminomadic owners for prestige—no economic use; but used in USA in crossbreeding for rodeo stock.

Africander

Farmed and originated: S Africa. Coat: usually solid red; also light red, yellow, gray. Horns: extended outward with forward curve and upturn; oval in cross section, on average more than 20in. Hump medium; dewlap quite large; sheath and umbilical fold not prominent. Features: long head; characteristic horns. wt: cow 1,102lb, bull 1,543lb ht: cow 53in, bull 59in. Beef and draft breed; poor milkers.

Nguni

Farmed and originated: E coast of southern Africa—Zululand (S Africa), Swaziland and Mozambique. Local races: Landim (Mozambique), Bavenda (Transvaal). Coat: whole, brindle or pied white, black, brown, red, tan. Horns: rather thin, wide, upswept, often lyre-shaped. Hump fairly well developed in bull, small or absent in cow; dewlap, sheath and umbilical fold little developed. wt: cow 660lb, bull 880lb. ht: cow 47in, bull 55in. Draft, beef and dairy breed.

Santa Gertrudis

Farmed and originated: Texas. Coat: solid cherry-red, perhaps some white spots on underside. Horns: thick, rather short, frequently bending down or forward; polled strain developed. Hump: medium (bull), not visible (cow). Dewlap, sheath and umbilical fold quite pronounced. Features: heat-tolerant, insect- and disease-resistant, massively built. wt: cow 1,322lb, bull 2,204lb; ht: cow 55in, bull 67in. Beef breed.

Bali cattle

Farmed and originated: Indonesia. Domesticated banteng. Coat: red with black stripe along back, or black (mature bull). Both sexes have white areas on hindquarters and belly and on legs from hooves to hocks. Horns: medium length, quite thick, swept upward, then inward at tops (bull); much finer in cow. Hump quite pronounced (see p44): dewlap medium; sheath and umbilical fold not prominent. Features; body wide and deep. wt: cow 573lb, bull 838lb. ht: cow 43in, bull 51in. Draft and beef breed.

Mithun

Mithun or Gayal

Hills to N and S of Valley of Assam. Domesticated gaur. Coat: black (bull), brown (cow); white stockings. Horns: broader, less curved than in gaur. Hump quite pronounced, different structure from that of zebu and sanga cattle (see p44). Dewlap prominent; sheath and umbilical fold not pronounced. Features; resembles gaur. wt: bull 880–1,102lb. ht: cow 51in, bull 55in. Kept for sacrifices.

◄▼ **Zebu, sanga and other breeds.** (**1**) Madura bull racing (Indonesia, from cross 1,500 years ago between banteng and cattle: beef and racing). (**2**) Ankole cow (East African sanga: ownership of ritual, not economic, importance). (**3**) Nguni cow (southern African sanga: draft, beef and dairy). (**4**) Santa Gertrudis bull (American sanga: beef). (**5**) Kankrej bull (Indian zebu: draft and dairy). (**6**) Brahman bull (American zebu: beef). (**7**) Gir bull (Indian zebu: draft and dairy). (**8**) Red Sindhi (Pakistani zebu: dairy).

Milk from Contented Cows

Operation of a dairy farm

Modern dairying follows the traditions of Europe where many dairy breeds and practices were developed. New Zealand sets the pace for low-cost production with just a milking parlor and cows unhoused on lush pasture. Israel, which milks its cows three times daily, has the highest production per cow of any nation.

Over time, capital investments for cow comfort and sanitary requirements have increased markedly. Labor-saving practices have been developed to reduce the drudgery of dairy farming. Many of the top producing cows continue to be housed and milked in labor-intensive stall barns. For these stall barns there now are battery-operated silage carts, portable straw choppers, automatic detaching milking machines with low milk lines, and mechanized manure handling.

Still these barns are not perfect and, where possible, many dairy farmers let their cows out of their barns in the middle of the day. On pleasant days, cows are more likely to groom themselves and each other, exercise their limbs, "sun" themselves, and show receptive behavior.

Economic pressures have forced many dairy farmers into programs of close confinement and intensification. Some animals are kept on concrete throughout their productive lives. These trends were accepted at first by producers and scientists without adequate data on the long-term effects of such confinement, but concern has since been expressed over the effects of the environment upon comfort, well-being, behavior, reproduction, udder health and feet and leg structure.

The dairy cow has been called "the foster mother of the human race." A relationship develops between the milker and the cow which is a vital part of the milk extraction process. As machine milking took over from hand milking this relationship was considered by many to have diminished. After her calf is removed, the cow is milked with a minimum of manual stimulation in highly automated surroundings.

Caretakers in high-producing herds are aware of the importance of such changes. For as long as cows have been milked there has been an art of cow care that results in more milk from healthier, more contented cows. It has been recognized that the dairy cow's productivity can be adversely affected by discomfort or uneasiness. Behavioral indications of adjustment to the environment are always useful signs of whether the environment needs to be improved. In some cases, the way animals behave is the only clue that stress is present.

Many dairymen allow their cows to develop their own individual personalities as long as it does not require special care and treatment. Mass handling of cows dictates that individual cows fit into the system rather than the system conforming to the habits of the cow. The slow milker, the kicker, the boss cow, the explorer and the finicky eater usually are removed from larger herds, regardless of pedigree.

Although concern is expressed from time to time about temperament and behavioral problems, most attempts at reinforcing correct behavior and disciplining improper behavior have been successful. One dairy study showed that behavior is a reason for disposal in less than 1 percent of cases. Other categories include: udder problems and mastitis—23 percent; low yield—4 percent; reproductive disorders and diseases—36 percent; digestive problems—11 percent; metabolic problems—7 percent; anatomical problems (feet, legs and skeleton)—11 percent. The cows culled for behavior represent the truly wild ones which would not conform to training and management.

As creatures of habit, gentle dairy animals may be excited into rebellion by the use of unnecessarily severe methods of handling and restraint. Attempts to force an animal to do something it does not want to

▶ **Designed for sound management of the herd,** ABOVE this model dairy farm provides individual stalls and outdoor feeding. The farm grows, stores and mixes its own corn and alfalfa for feed and conducts a breeding program. The model is based on Hoard's Dairyman Farm in Wisconsin, established in 1899.

▶ **Milking in the round.** BELOW As each cow steps onto this rotating platform at a dairy farm in Scotland, her number is keyed into a computer. The computer signals a feed hopper and chute to give the cow her individually required feed supplements. The human operator washes her udder and puts on the milking machine teat cups. As the platform makes one circuit the cow is milked out and the teat cups are removed automatically before she steps away. Both rotary and stationary milking systems can include: the abreast parlor, as here, where cows stand side by side with their backs to the milking-machine operator (see also p16); the tandem parlor, where cows stand in lines one behind the other with their sides to the operator; and the herringbone parlor, where cows walk into rows of stalls which leave them standing at a 45° angle to the operator. In all of these systems, milk can be piped directly through the milking machines into refrigerated bulk storage tanks.

◀ **Milk yield** during lactation period of a typical Friesian cow. Daily yield reaches a maximum a few weeks after calving (up to 95 US pints—45 litres—in some cows). In a natural wild state a cow would annually produce less than (2,750 US pints 1,300 litres), but selective breeding and intensive farming can produce yields ten times that amount.

▼ **The milk year** starts with the birth of a calf, who stays with the mother for a few days. The cow then rejoins the milking herd. The cow is fertilized, usually by artificial insemination, about 85 days after the birth of the previous calf. Calving may be planned for the spring, which pushes the cycle around 5–6 months. Fall calving is more expensive as it relies more on winter feed which is costly. On the other hand, the maximum milk yield occurs in the winter when prices are better.

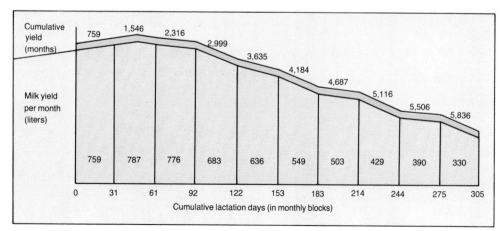

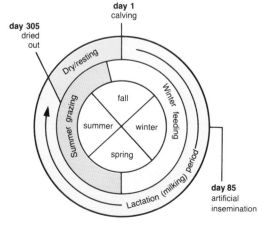

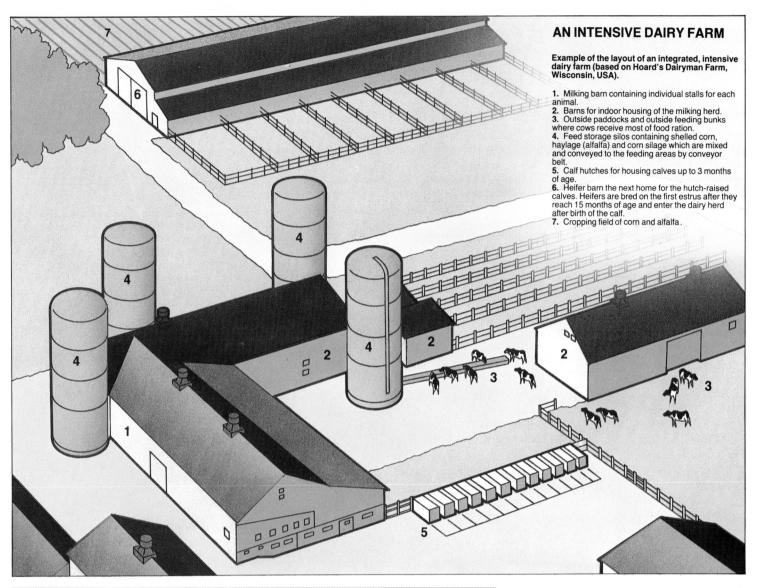

AN INTENSIVE DAIRY FARM

Example of the layout of an integrated, intensive dairy farm (based on Hoard's Dairyman Farm, Wisconsin, USA).

1. Milking barn containing individual stalls for each animal.
2. Barns for indoor housing of the milking herd.
3. Outside paddocks and outside feeding bunks where cows receive most of food ration.
4. Feed storage silos containing shelled corn, haylage (alfalfa) and corn silage which are mixed and conveyed to the feeding areas by conveyor belt.
5. Calf hutches for housing calves up to 3 months of age.
6. Heifer barn the next home for the hutch-raised calves. Heifers are bred on the first estrus after they reach 15 months of age and enter the dairy herd after birth of the calf.
7. Cropping field of corn and alfalfa.

do often end in failure and can cause the animal to become confused, disorientated, frightened or upset. Handling livestock requires that they be outsmarted rather than outfought and that they be outwaited rather than hurried. Designing of a dairy system for welfare is only part of the solution. The caretaker's attitude and behavior are of the greatest importance.

Considerable self-stimulation and "inwardness" occur in cattle due to the rumination process. During rumination, cows appear relaxed with their heads down and their eyelids lowered. Also, through cud-chewing and mutual and self-grooming, aggression is reduced and there is little or no boredom.

Management developments which have improved the comfort and well-being of dairy cattle include raising calves in individual pens or hutches, providing exercise prior to calving, grooving or roughening any concrete flooring, making use of pasture or earthen exercise lots and eliminating slats. Individual stalls called cubicles or free stalls have proved better than loose housing. Dairy cattle thrive best when they are kept cool, free from flies and pests and provided with bedding.   JLA

WATER BUFFALO

Bubalus arnee ("B. bubalis")

One of 4 species of genus *Bubalus*
Order: Artiodactyla.
Suborder: Ruminantia.
Family: Bovidae.
Three breed groups: swamp, river and *desi*.
Distribution: swamp buffalo—eastern half of Asia from India to Taiwan, Papua New Guinea (in Philippines known as carabao); river buffalo—western half of Asia, Asia Minor, Egypt, eastern Mediterranean (Bulgaria, Yugoslavia, USSR, Greece and Hungary); *desi* (unimproved)—India and Pakistan. Feral in N Australia, Brazil, Guam, Java and Sri Lanka. Total world population (farmed and feral): 130 million.

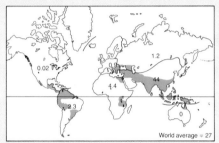

Figures show number of farmed Water buffalo, by region, relative to 1,000 head of human population (FAO data 1982).

Size: very variable; height at withers, 40–66in (100–170cm); weight: 550–1,300lb (250–600kg) perhaps to 2,200lb (1,000kg). Feral bull swamp buffalo in Northern Territory of Australia may reach 2,650lb (1,200kg).

Color: dark gray or slate black; swamp buffalo and the Surti breed of river buffalo with white chevrons below jaw and on chest; locally (eg Thailand) all-white animals can be common.

Horns: main feature used to distinguish breeds; no polled breeds described; massive, growing outward from the head in semicircle (swamp buffalo); set closely to head, may be downswept (river buffalo).

Other features: large, not particularly drooping ears; broad, rather flat face and muzzle; head may be domed between horns.

Diet: appear to eat a wider range of plants than cattle.

Metabolism: more sensitive than cattle to extremes of heat and cold.

Dental formula: I 0/4, C 0/0, P 3/3, M 3/3 = 32.

Breeding: adequately fed females mature at 14–16 months, males at 18 months, and breed all year round. Artificial insemination can be performed; receptivity difficult to detect; it lasts 24 hours and occurs about every 21 days until pregnancy achieved. Gestation 300–334 days; may calve every 14 months.

Longevity: normally 20 years, up to 40.

HUNDREDS of millions of people have very close personal attachments to individual animals and will claim that the animal returns the affection. Some species seem particularly disposed to this kind of relationship; the dog, the horse and the Water buffalo. Docile enough to be handled by small children, after a working life of 20 years a Water buffalo may (in Buddhist countries particularly) be pensioned off to live in retirement as a member of the family.

The species had been domesticated in the Indus Valley by 2500BC, in what is now Pakistan; there is no archaeological evidence yet, but it was probably domesticated even before that in the rice growing areas of southern China or Indochina. Water buffalos were unknown to the ancient Egyptians, Romans and Greeks but in the early Middle Ages they were introduced to the eastern Mediterranean. Since about the 7th century AD they have been common draft and milk animals in Italy and southern Europe. Though in prehistoric times cattle were taken to Africa from the Middle and Near East, Water buffalos do not ever seem to have become established there except in Egypt. In 1807 Napoleon brought them from Italy to the Landes district of southwestern France where they did exceedingly well. They soon met their Waterloo, however; after Napoleon's fall they were quickly exterminated.

Water buffalo are divided into swamp type (particularly suited for draft and meat), river type (principally milk, males being used for draft) and *desi* (the nondescript unspecialized, "mongrel" type of the Indian subcontinent). Only the river type of India can be subdivided into breed groups (Murrah and Gujarat) and further into "local breeds." Mediterranean and Egyptian Water buffalos have developed separately from the other river type populations.

Half the world's human population depends on rice and three-quarters of the world's Water buffalo live in traditional rice growing areas. The Water buffalo is the ideal animal to help in cultivating it and is aptly described as the "living tractor of the East." Its large hooves enable it to walk securely through mud. One or two buffalos individually yoked and controlled by a rein attached to a nose plug are used to plow the paddy. Then it is flooded and the buffalos work belly deep, harrowing the soil to the correct consistency. Carting and threshing the harvested rice is also done by the farm buffalo which has an average working year of 60–100 days, after which it may be released to forage for itself until the next season. On average a buffalo can plow 0.62 acre (0.25ha) per day and can cope with 7.5 acres (3ha) of paddy.

In the third world millions of farms are less than 12.5 acres (5ha) and the most valuable possession of the small farmer is, very often, a pair of buffalos or even a lone animal. The buffalo has limitations, of course; it will work at most six hours a day,

▼ **In danger of acute distress** if left in direct sun for even a few hours, Water buffalos have many fewer sweat glands and a less efficient system for heat regulation than cattle. They must be regularly sluiced and sprayed or allowed periods of rest in the shade or a wallow. River breeds prefer to wallow in the clear water of streams and ponds but swamp buffalos such as these in Sri Lanka like to wallow in mud, which they stir with their massive horns.

it cannot tolerate long exposure to the sun and has to wallow in water or mud in the heat of the day and at intervals while working, to be able to cool off. If access to water is denied them they will seek out water at the first opportunity regardless of what they are pulling or carrying, with potentially disastrous consequences. A wallowing Water buffalo may be a picture of idyllic content-

ment but its life is a hard one; buffalos are susceptible to the same diseases as cattle (except that ticks and flies are usually less of a problem, presumably because of the wallowing habit) and buffalo calves die in large numbers, very often because their mothers' milk is sold for cash.

The vast majority are not at all pampered, having to subsist on poor forages like straw

and other harvest leftovers (which, compared to cattle, buffalos appear to digest more efficiently) and having to work hardest at the start of the rainy season, preparing the fields, at a time when the ground is hardest, temperature and humidity at their highest and food supplies most limited in quantity and quality. Usually the harnesses of their implements are inefficient and prevent the animal from exerting its full strength. Poor feeding means that their reproductive potential is usually not fulfilled. Opportunities for selective breeding of Water buffalo are being lost because the quick-growing bulls are more likely to be prized as draft animals and castrated rather than used to upgrade stock.

In Taiwan, India and Pakistan buffalos are used for road haulage and may also haul wagons on narrow-gauge railway lines. A single buffalo can pull a cart with a 220lb (100kg) load at 2mi (3km) per hour while a pair of entire males can pull 2 tons a distance of 15–20 mi (25–30km) in a day. For larger loads, teams of up to eight buffalos or mixed teams of buffalos and oxen may be used. These traction animals are shod with pieces of car tires or pads of woven straw. Other uses of buffalos include sugar cane pressing, raising water and forestry work.

The Water buffalo is overwhelmingly a working animal of the village, but it does have tremendous potential for development and selective enhancement of its biology. In India special milking breeds have been developed and 70 percent of India's milk comes from buffalos. In Egypt (where it is the most important domestic animal) milking buffalos usually provide more milk, 1,500–1,750lb (680–800kg) in a lactation of about 285 days, than local cattle (800–1,100lb—360–500kg). Buffalo milk is rich: 17.5lb (8kg) of cow's milk is needed to make 2.2lb (1kg) of cheese, but only 11lb (5kg) of buffalo milk. It contains more than 16 percent total solids (cow's milk 12–14 percent). About 6 percent of the world's milk supply comes from Water buffalos and it could well be developed further as a tropical dairy animal. Thousands of small herds are kept in the streets. Similarly, it has potential as a meat animal though it will

▲▶ **Toiler of forest and field.** ABOVE A pair of entire male buffalos can pull a 2-ton load 15–20mi (25–30km) in a day. RIGHT Buffalos plow a rice paddy in Java. The Water buffalo is a placid, easily trained beast of burden. Used not only to haul and plow, it powers mills for grain, oilseed and sugar cane, threshes rice sheaves by trampling, raises water from wells and carries packs and riders.

▶ **World importance of Water buffalos.** BELOW Numbers are highest in India, China and Pakistan. The swamp type is most common in southern India and eastwards through China to the Philippines. Used mainly for draft, especially for plowing rice paddies, it is also milked but provides a low yield. In northern India, the river type predominates and there are recognized breeds, developed especially for milk. However, most Indian buffalos are *desi* (of no particular breed) and often intermediate, especially in the south, between the swamp and river types. *Desi* are mainly draft animals used for plowing and hauling. Middle Eastern river-type breeds have been developed as dual-purpose meat and milk breeds. Water buffalos convert poor feed efficiently; the potential for meat production from male calves of dairy animals is very high.

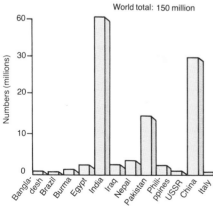
Water buffalo numbers in selected countries 1979
World total: 150 million

do not provide. Buffalo beef is lean, a selling point in many markets, and buffalo milk is particularly good for cheese (Mozzarella cheese comes from the Italian herds.) However the real significance of the animal is in the traditional agriculture of the third world where its supremacy will remain unchallenged for the foreseeable future.

There are still a few wild Water buffalos in the inaccessible grass jungles of Nepal and in some wildlife refuges in India. Feral herds graze on three continents: a quarter of a million around Darwin (Australia), with small herds on Guam and Java in the western Pacific, on Ilha de Marajo at the mouth of the Amazon, in Sri Lanka and in many other places where domestic animals have escaped.

The three close living relatives of the Water buffalo (the anoas, *Bubalus quarlesi* and *B. depressicornis*, of Sulawesi, and the tamaraw, *B. mindorensis*, of Mindoro in the Philippines) do not appear ever to have been domesticated. Not much bigger than goats, they are in effect miniature Water buffalos, living a secretive life in forests in small social groups. In contrast to these species and to the other tropical wild cattle species (the gaur, banteng and kouprey), the Water buffalo seems willing to merge its family groups into big herds which may number from 30–50. This may be linked with its greater preference for open, grassy habitat.

The basic building block of Water buffalo social life in the wild is the family group or "clan." Female calves stay with their mothers for life and their female calves also become permanent members of the group. Bull calves leave before they are three years old and live in bull groups, usually up to ten in number, associating with cow groups at mating time. In all their daily activities, Water buffalo copy each other, moving together from the grazing area to the lying area or to and from the wallow, always defecating in the same places, always rubbing on the same trees. All this points to a social organization based on personal knowledge of one's closest relatives. It is hardly surprising that such an intensely social animal can integrate so well into human societies and activities.

There is a huge amount to be learned about the farmed Water buffalo, and great rewards could be reaped from the application of research. Such work would lead directly to man's benefit, as third world agriculture depends so utterly on this animal, and western farming and ranching could gain if the Water buffalo became recognized as a valuable meat and milk animal.   SJGH

probably not respond to intensive feeding in such a cost-effective way as beef cattle.

It is only in this century that the Water buffalo has been adopted outside Asia and the Mediterranean. Stock mainly from Italy and India have done very well in Brazil and elsewhere in South America. There are small herds in Papua New Guinea, and experimental herds in the United States and Africa. It has a role to play in modern agriculture; in many habitats it can out-perform cattle, as a dairy or range beef animal, either by being better adapted to the environment or because of valuable products which cattle

YAK, LLAMA, ALPACA

 Yak

Llama
Alpaca

Yak
Bos mutus ("*B. grunniens*")
Family: Bovidae.
Distribution: farmed in Bhutan, Mongolia, Nepal, Kashmir, Tibet, China, USSR; wild in scattered localities on alpine tundra and ice deserts of the Tibetan Plateau at altitudes of 13,000–20,000ft (4,000–6,000m).

 Size: bull, height at withers 45–52.5in (115–135cm), weight 660–1,210lb (300–550kg): cow, height at withers 39–49in (100–124cm); weight 396–770lb (180–350kg).

Coat: shaggy fringes of coarse hair about body with dense undercoat of soft hair. Color blackish brown with white around muzzle.
Horns: wide apart; length in bull up to 38in (97cm), in cow 20in (51cm).
Gestation: 340 days.

Llama
Lama glama
Family: Camelidae.
Two breeds: Chaku, Ccara.
Distribution: Andes of C Peru, W Bolivia, NE Chile, NW Argentina, Alpine grassland and shrubland at 7,546–13,123ft (2,300–4,000m). Population: 3.7 million.

 Size: head-body length 47–88in (120–225cm), height at withers 43–47in (109–119cm); weight 285–340lb (130–155kg).

Coat: uniform or multicolored white, brown, gray to black.
Gestation: 348–368 days.

Alpaca
Lama pacos
Family: Camelidae.
Two breeds: Huacaya, Suri.
Distribution: Andes of C Peru to W Bolivia. Alpine grassland, meadows and marshes at 14,430–15,750ft. Population 3.3 million.

 Size: head-body length 47–88in (120–225cm), height at withers 37–41in (94–104cm); weight 120–140lb (55–65kg).

Coat: uniform or multicolored white, brown, gray to black; hair longer than llama's.
Gestation: 342–345 days.

Two top-of-the-world cultures have developed domesticated animals which are uniquely their own: the Tibetan civilization depends heavily on the yak; and, since long before the time of the Incas, Indians of the high Andes have reared and exploited the llama.

Wild **yaks** are used for their meat and hide. When domesticated, they serve as mounts or pack animals, carrying loads of 330lb (150kg) over steep mountain paths at elevations above 6,500ft (2,000m); even when poorly fed they can carry 110–150lb (50–75kg) a distance of 8–10mi (13–16km) per day for months and still remain in good condition. Yak milk is aromatic with a high nutritive value. The butterfat content is about 6.5 percent. Milk production is about 423qt (400l) per year. Yaks are shorn once annually with a yield of about 6.5lb (3kg) of coarse wool that is spun into essentials. Where there are no trees or bushes in the Tibetan highlands, dried yak manure is used for fuel.

In its domesticated form, the yak is considered to be a cross between the wild yak and some form of domesticated cattle; it bears a resemblance to the extinct wild aurochs. It is also like a bison, with its rounded forehead, shoulder hump and shaggy coat. Yaks, like bison, have 14 pairs of ribs instead of the 13 in cattle, and like bison they do not low but make groaning sounds which resemble the grunting of pigs. Unlike bison, yaks break through the snow with their horns to get at food. Bison and beefalo (cattle crossed with bison) graze by swinging their heads from side to side, and

their muzzles have been known to become so sore from crusted snow that starvation has resulted.

It is an advantage to their owners that yaks can live on the wiry grasses which grow in their harsh homeland. A disadvantage is that they will not eat grain and must often be moved great distances when pastures are depleted.

Tibetan nomads have been successfully crossbreeding yaks with zebu cattle (see pp44–47) for more than 500 years. The cross between Chinese yellow cattle and yak produces a hybrid known as the Pien Niu. The protein content of the meat from these crosses is comparable to that of the beefalo. The milk, greater in yield, is about 50 percent richer in butterfat than the milk from the zebu breeds used. They also make excellent draft animals at moderate altitudes. Although they are not as strong as purebred yaks and the males are sterile, crosses have the advantage that they can be made to pull a plow.

In Britain, a Lincolnshire farmer aiming at an animal that will produce lean meat on a scrubland diet has produced a "yakow" by crossing a Highland cow with a yak. Unsuccessful attempts have been made for more than 50 years to cross wild yak bulls with North American cows to produce an animal which can feel at home in bitter Saskatchewan winters in Canada. Domesticated yaks have not been used for these experiments because they have been thought too small to sire a large breed.

Domestication of the **llama** is thought to have first taken place from about 7,000 years ago in the region of Lake Titicaca on the present-day borders of Bolivia and Peru, or on the Junin Plateau in Peru. The pre-Incan *Qollas* culture on the highlands west of Lake Titicaca were well known throughout the Andes as experienced animal breeders, who controlled vast herds of llamas and alpacas that later became the nucleus of the Inca state herds.

The Inca Empire's culture and economy revolved around the llama, providing the primary means of transportation and carrying goods. Hundreds of thousands were used by the state for commerce and mining. In addition to the llama's use as a beast of burden and the alpaca's use for the production of fine wool, both have traditionally been used for meat, hides, fuel (dung), medicines, and religious ceremonies. Neither were commonly used for riding nor pulling of plows or carts.

The llama is tolerant of the high altitudes of the Andes because of its specialized blood

▲ **In the early morning snow,** at 15,750ft (4,800m), Bolivian llama drivers prepare their animals for a day of high-altitude transport. With blood that has a higher oxygen-carrying capacity than other domesticated animals', llamas have been the chief means of moving goods through the high Andes since before the days of the Incas. As roads and railways reach more and more remote areas, their numbers (still about 3.7 million) are declining.

◄ **Saddling up their yaks,** these Tibetan herdsmen break camp beneath the icy peaks of Kantega (22,340ft—6,809m). Yaks will not eat grain and must be moved long distances when pasture becomes depleted. They serve as mounts and high-altitude pack animals, as well as providing rich dairy products. Because of religious prohibitions their meat is seldom eaten in Tibet.

▶ **Two breeds of alpaca,** OVERLEAF the Huacaya standing, with shorter wool fibers, and the Suri, with long, straight fibers in waves. The color of both breeds varies from white to black and brown. Peru exports about 2,400 tons of alpaca wool annually.

hemoglobin which has a higher oxygen carrying capacity than that of other animals. Today, the docile llama is used mainly as a pack animal, carrying loads of 55–130lb (25–60kg) for 9–18mi (15–30km) per day, for up to 5 days before needing a rest. However, as more road and rail transport comes into service in this rugged country, llama numbers are beginning to decline. Bolivia has 70 percent of the world's 3.7 million population.

The 3.3 million **alpacas** are raised for their long, fine wool ranging in color from black-brown to white. Amost all (91 percent) are in Peru. Due to the value of their wool their numbers are increasing.

According to a long-held view, the llama and its close relative the alpaca are both domesticated forms of the wild guanaco (*Lama guanicöe*), although this animal shows differences in behavior and the growth of its incisor teeth. A more recent view is that the alpaca has resulted from crossbreeding either the llama or the guanaco with a fourth lamoid, the wild vicuna (*Vicugna vicugna*). All four groups have the same number of chromosomes (74) and can interbreed to produce fully fertile offspring.

Llamas and alpacas mix freely and out of this association come two hybrid animals, each possessing certain of the characteristics of both animals. These animals are the *huarizo*, born of a llama father and an alpaca mother, and *misti*, the progeny of an alpaca father and a llama mother. In some remote regions, the *huarizo* is used as a beast of burden. The fleeces of these hybrids are not as fine in either texture or quality as the alpaca's.

Like their close relative the camel, llamas spit when they become excited or aggressive. They direct regurgitated, partly digested food at each other or at human targets. Llamas are territorial and in a pasture will eat the perimeters of their territory, leaving the middle intact.

Large-bodied, well-nourished llamas may breed before they are one year old, but most females first breed at two years. A single offspring, born from a standing position, is mobile and following its mother within 15–30 minutes. The female comes into heat again 24 hours after giving birth, but usually does not mate for two weeks. Non-breeding males are castrated.

The Suri breed of alpaca is the best known, for both its long straight and its wavy wool fibers. A second breed is the Huacaya. The llama as well is known under two breed names, Chaku and Ccara.   JLA

SHEEP

"Ovis aries" (*O. ammon, O. musimon*)
Order: Artiodactyla.
Suborder: Ruminantia.
Family: Bovidae.
Number of breeds and strains: approximately 300.
Distribution: worldwide, especially between 45° and 55°N in Europe and Asia and 30° and 45°S in S America, Australia and New Zealand. Lesser numbers in India and Africa between 5° and 35°N. Total population worldwide is about 1.13 billion as well as wild populations of argali, bighorn, urial and mouflon.

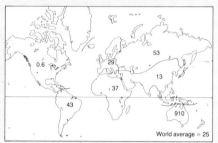

Number of sheep, by region, relative to 100 head of human population (FAO Data 1982).

Size: males, height at withers 17–33in, weight 55–330lb; females, height at withers 15–31in, weight 44–230lb.

Distinguished from goats by tail (hanging down), inguinal glands, sub-orbital tear glands and glands between hooves.

Horns: with downward curve or tight spiral, particularly in males, or straight and growing almost vertically or horizontally: generally 2; in some breeds only males horned and in others, both sexes polled (hornless).

Temperature 101.3–105.0°F (38.4–41.0°C)

Respiration rate: 12–20/min (tends to be higher in smaller animals and in pregnancy).

Heart rate: 70–80/min.

Dental formula: I0/3, C0/1, P3/3, M3/3 = 32.

Longevity: rarely survive until old age; culled when they lose their teeth and productivity declines. Ewes of 10 years will still produce lambs but have difficulty in rearing them.

Breeding: ewes from breeds developed in high latitudes mate during the fall. Rams produce semen all the year, though semen production and fertility tend to be lower in summer. Breeds from other latitudes aseasonal but may have peaks of mating with variations in feed. If the ewe does not become pregnant she will come into heat every 16.5–17.5 days for 30–36 hours for a period of about five months. Pregnancy varies from 143–159 days.

► **A wild ancestor of domestic sheep,** the mouflon's natural range is in Asia Minor.

► **Climbing the Pyrenees,** OPPOSITE this flock in the Béarn region of France is part of the 10 percent of French sheep still kept in a "transhumant" system, moving great distances between winter and summer pasture.

WILD sheep and goats evolved in the dry, mountainous region of southwest and central Asia some 2.5 million years ago, in the Pleistocene period. The first ancestors were as large as cattle. At the end of the ice ages true sheep first appeared, spreading rapidly into areas left by the retreating ice and reaching North America and northern Africa by the late Pleistocene period.

It is a matter of controversy how much each of the different types of wild sheep have contributed to the genetic make-up of the domesticated species. The bighorn (*Ovis canadensis*), which was never domesticated has not contributed anything. The urial (*O. orientalis*) occurs wild in the area where domesticated sheep first appeared, but only the mouflon (*O. musimon*) has the same number of chromosomes as domesticated sheep (54). The urial has 58 and the argali has 56. The argali (*O. ammon*), however, could be a joint ancestor with the mouflon: mouflon-argali hybrids have 55 chromosomes but produce ova (eggs) with 27, the right number to produce offspring with 54 chromosomes.

Domestication of Sheep

Sheep and goats were domesticated after dogs and before cattle, pigs and horses. Domestication was carried out during the Mesolithic period (10,000–8,000 years ago) in the area of grain farming in the fertile crescent that extended from present-day Israel through Lebanon and Syria across southern Turkey to the Iraq-Iran border. There is some speculation that goats may have been domesticated first, as they may have been more useful to clear trees and shrubs for arable farming, but it is impossible to be certain, because of the difficulty of distinguishing the bones of goats from those of sheep. Certainly, by 9000BC sheep had been domesticated; large numbers of sheep bones dating from this time have been found at a site in northeast Iraq.

The first domesticated sheep were smaller than wild sheep. The bones, particularly in the legs, remained fine but were considerably shortened and tails became longer. The main changes with domestication were, however, in the fleece; wild sheep have a heavy outer coat of kemp with an average fiber diameter of 0.004–0.0056in (100–140 micrometers) and a small, barely visible undercoat of wool with a diameter of 0.0006–0.0008in (15–20 micrometers), finer than modern fine wools. The fleece was shed in spring; this still happens in some primitive breeds, like the Soay and Mongolian Fat-tail sheep, and enables the fine wool to be collected by combing, as the coarse fibers are retained longer. Sheep now produce three types of fiber: wool (see pp19–20), kemp and hair, which is an intermediate fiber.

When tool making had developed to the stage where sharp knives were available, a

World Importance of Sheep

Sheep and goats together account for 6 percent of world production of meat, 3 percent of milk production and 5 percent of the fiber used in textile manufacture. Large exports of meat and wool from Australia and New Zealand to northern Europe, particularly Britain, have occurred for over a century and recently Japan has become a major importer.

About 10 percent of sheep meat production is traded internationally. The largest exporter, New Zealand, still sends about a third of its total production to Britain, and Australia sends a similar proportion to Japan. There is a large movement of live sheep. About 14 million are exported, almost entirely to the oil-producing countries of the Middle East. A third are transported by sea from Australia, a slightly smaller number from Afghanistan and the remainder from eastern Europe and East Africa (Somalia and Sudan). This trade reaches a peak to provide animals for sacrificial slaughter at the end of the major fast, Ramadan.

The major wool-producing countries are mainly in the Southern Hemisphere and most of this wool, from Merino or longwool-Merino crossbred sheep, is exported unprocessed for manufacture in the Northern Hemisphere, where Japan, the USSR, Britain, France, Italy and West Germany are the largest importers. About 25 percent of the world's wool is produced in Australia, all of it exported. The USSR, the second largest producer, is also a major importer. The other major producers are China, New Zealand, Argentina, South Africa and Uruguay.

*Production of sheep meat only.

	Population (thousands)	Sheep and Goat Meat (1,000 tons)	Wool (1,000 tons)	Milk (1,000 tons)
Africa	184,301	1,297	206.9	697
South Africa	31,650	158	111.4	0
Ethiopia	23,330	132	12.3	58
Sudan	18,125	142	15.6	127
Morocco	14,840	64	13.1	21
Algeria	13,600	76	21.3	164
Nigeria	12,000	167	na	0
Somalia	10,200	86	na	0
North & Central America	22,966	214	60.2	0
USA	12,936	154*	50.3	0
Mexico	7,990	35	8.6	0
South America	105,165	328	317.5	35
Argentina	30,000	119*	165.1	0
Uruguay	20,429	38	71.1	0
Brazil	18,000	52	30.5	28
Peru	14,671	31	13.0	0
Bolivia	8,900	27	8.9	0
Chile	6,185	21	21.6	0
Asia	335,035	2,835	439.7	3,579
China	105,200	761	179.5	495
Turkey	48,630	442	62.2	1,200
India	41,500	404	36.0	0
Iran	34,377	233	16.0	705
Pakistan	24,468	340	46.2	39
Afghanistan	20,000	139	22.9	225
Syria	11,738	94	16.6	476
Iraq	11,650	51	18.2	130
Europe	137,109	1,269	278.8	3,500
UK	32,282	255*	52.2	0
Rumania	15,865	80	40.0	340
Spain	14,887	140	21.3	209
France	12,980	175*	22.2	1,082
Bulgaria	10,433	88	34.5	308
Greece	7,920	121*	8.1	580
Australasia	204,601	945	1,085.4	0
Australia	133,396	325*	700.4	0
New Zealand	71,200	620*	385.0	0
USSR	141,573	800	454.0	100
Total:	1,130,751	8,099	2,842.5	7,910

further stage in domestication was possible. Sheep that did not molt naturally and could have their fleece shorn off were an advantage and were selected. The last major step in domestication was selection against pigmented fibers, to give white wool that can be dyed to any color. By 3000BC a wide variety of sheep types were depicted in carved friezes and recorded on tablets in Babylon, the Land of Wool.

The Spread of Domesticated Sheep
From southwest Asia domesticated sheep were taken by 6000BC into all western Asia, southeastern Europe and the islands of Cyprus and Crete. In the next thousand years they spread into North Africa, along the north side of the Mediterranean and, following the Danube route, into central Europe. By 4000BC, when they had been transported to the British Isles and reached east as far as China, the distribution had reached limits which were not broken until the discovery of the Americas and Australasia. In the middle of the 16th century the Spanish took sheep, mainly of the Churra breed, to South America and to the south of North America, and, about a century later, English and Dutch settlers brought sheep to the eastern coast of North America. The final movement of sheep was to Australia. In 1788 it was settled as a convict colony and African and Indian Hair sheep were imported from South Africa and Bengal for fresh meat. Later, in 1797, Merinos were brought from South Africa to found the flocks on which the whole development of the continent rested.

Role of Sheep Past and Present
Wild sheep provided hunters with meat and fat for food, pelts for clothing, and bones and horns for tools and implements. With domestication the number of products increased and no other animal provided so many human needs. The most important produce, at least until recent times, was wool, but as wild sheep and the first domesticated sheep had only a very small amount under their coarse winter coat, it seems unlikely they were originally domesticated for wool.

▶ **Probably predating wool** ABOVE as an agricultural product, sheep's milk is twice as rich in fat as cow's milk and contains more protein. These ewes are being milked in the Osseau Valley in the French Pyrenees. Well-known sheep's milk cheeses include Ricotta, an Italian whey cheese, and Roquefort, a blue cheese matured in cool caves of the Aveyron region of central France.

▶ **Feed supplements** MIDDLE are required by ewes for about two-thirds of the year. Only in summer, when they are neither pregnant nor nursing and pastures are at their best, can grazing provide adequate nutrition. During the rest of the year they may be fed root crops, hay, rolled oats or prepared pellets and still spend longer hours grazing.

▶ **Making a leisured progress,** BELOW Australian Merinos move along an outback road to greener pastures. Australia is by far the leading world producer and exporter of wool, but as demand for wool has fallen, so has the number of Australian sheep. Merinos can efficiently turn sparse grazing into wool, but meat breeds are not successful on the same pastures, and farmers moving from wool to meat production have tended to change over to cattle. Sheepmeat is exported from the moister regions of Australia.

As the fleece became modified the use of sheep for meat almost certainly declined, and animals were only slaughtered at festivals or religious feasts or when numbers seriously exceeded supplies of pasture. This is still so with nomadic flocks of Asia and North Africa whose owners take the wool and milk of their sheep and, occasionally, the blood. Among some of these people the size of flock is important: it not only confers status but is also an important investment and form of insurance.

Spinning and weaving were developed by Neolithic times (8,000 years ago), but felt, made by repeatedly compressing and wetting wool, almost certainly is a more ancient fabric than woven cloth. Felt tents (yurts), cloaks and boots were a vital necessity in the bitter winters of central Asia and the availability of wool was critical for the colonization by man of these areas.

With domestication came milking, and the sheep is still considered foremost a dairy animal in many Mediterranean, Middle Eastern and Asiatic countries, where cattle production is difficult or impossible. The rich milk is rarely drunk. It is transformed into soured or curdled products such as yogurt and cheese to be stored.

It is more common to bleed cattle than sheep, but some nomads in Africa bleed sheep from a cut above the eye. Blood from slaughtered animals, mixed with grain or fat and cooked in the guts of sheep, is a very ancient food described in Homer's *Odyssey*.

Dung is an important product of sheep. Among nomadic people dried dung is a major source of fuel. Before chemical fertilizers, animal dung played a major part in crop production. There are many examples of systems in which sheep were kept for manure and both wool and meat were by-products. In China, the Hu breed produced manure for mulberry trees kept to feed silkworms. Until 50 years ago, keeping sheep on root crops was a part of many farming systems on poor quality land in Britain, and in European countries where shepherded flocks still exist, landowners pay to have the animals on their land at night.

Sheep meat is unique in having no taboo in any culture or religion, although one may be developing now in the fat-conscious western culture! All the organs can be eaten and many, like tails, testicles, eyes and brains are considered delicacies. The stomach and guts are important as casings for sausages, blood puddings and haggis made from minced meats and blood. Guts have provided strings for musical instruments and fishing lines. The use of the blind cecum of the sheep as a contraceptive is mentioned by both Casanova and Boswell, but the practice was already old in their time. The fat (tallow) is hard and from early Christian times was used to make candles.

Sheep skins in the past had many important uses both as pelts with the wool on, and after removal of the wool, by treating with alkalis. Pelts are still an important product for clothing in Asia. The highest quality pelts for the luxury market came from slaughtering 2–3-day-old lambs of breeds such as the Karakul from the USSR and the Tan and Hu from China. Colored skins from 4–6-month-old lambs of breeds like the Romanov in the USSR and the Gotland in Sweden are valuable for ordinary clothing. Pelts from other breeds including Merinos still have many industrial uses.

The skins, after removal of the wool, and tanning to make leather, often had special uses, including oilskin clothing and musical instruments (drums, the bag of the Scottish bagpipes). Until the 12th century, when paper was introduced by the Arabs, many sheep skins were used to make parchment. The skin, after removal of the wool or hair, is dried while stretched on a frame, shaved smooth and finally polished with pumice. Parchment, which is very durable, provided it is kept dry, gradually superseded papyrus in Greece and Rome. Used at first in scrolls, from the second century AD it was used in bound books. The folio size was dictated by

the size of a cured sheep skin folded once.

Although in the Middle Ages England and then Spain became noted for producing fine wool, the sheep were multi-purpose. They were milked and also provided dung, meat, pelts, hides, fat and bones. Breeds for special purposes only appeared with industrialization. In England this was the impetus for selecting breeds for meat production and led to improvements in the longwool breeds and later to the creation of the Down breeds. The demand for wool created by the new mills in Britain was a major factor in establishing fine wool production for the Merino sheep in Germany and then Australia. In Australia wool production became the sole purpose and even meat production was of minor importance.

In much of western Europe today sheep are only considered for meat production, and, as the value of wool has declined, have become single-purpose animals. In Asia and many areas of eastern Europe and the Mediterranean sheep are still kept in very ancient systems which have changed little for many hundreds of years. So there are still many triple-purpose (milk, meat, wool) breeds, with the importance of each product varying from region to region and with changing economic conditions.

Because they are not specialized for any one product these triple-purpose breeds vary greatly in type and origin. The majority provide carpet wool but there are triple-purpose Merino breeds in Italy and Spain.

In this century some specialized milk breeds have been selected. The Ostfriesisches Milchschaf from northern Germany is the best, with yields of 1.1–1.3lb (500–600g). It has been crossed in Israel with the indigenous Awassi to give, under good management conditions, yields of four to

5

7

6

▲ ▶ **Merino and finewool sheep breeds.**
(1) Tasmanian Merino ram, showing skin wrinkles which were bred into original unwrinkled stock. (2) Spanish Merino ewe, Serenna type. (3) Arles Merino ram, a leggy breed adapted to transhumance (migratory husbandry). (4) American Rambouillet ram, with better mutton carcass than other Merinos. (5) Ile de France ewe. (6) Textel ram, a breed producing very lean lamb. (7) Berrichonne du Cher ewe.

five times those of the original Awassi breed. In Italy, France, Greece and Spain modern methods of selection and progeny testing are being applied to a number of formerly triple-purpose breeds and will lead to new, improved dairy and meat breeds.

Merino Sheep

The name "Merino" first appears in Spain in the 15th century, but fine-wooled sheep were kept there from the 13th century. These sheep almost certainly derived originally from Asia Minor. Some reached Spain in Roman times, but it is believed that the Spanish Merino breed came from sheep introduced by the Beri-Merines, a Berber tribe that migrated to Spain from North Africa in the 12th century.

With the decline of British production in the 12th century, Spain developed a monopoly for fine wool. This wool came from enormous, migratory flocks of sheep summered in the mountains of the north of Spain and herded south to winter in the plains of Estremadura and Andalucia.

Associations of flockowners, the most famous being the Mesta of Castile, gained rights from the kings of Spain to grazing along the routes between the winter and summer pastures. Damage from the flocks brought constant strife with farmers and demands for these ancient rights to be controlled. With the decline in the power of flockowners' associations in the 17th century, wool production decreased.

The many different Merino breeds around the world are all derived from the Spanish Merino. Until the 18th century the flockowners' associations prevented the export of Merinos. A few may have been smuggled into France earlier but between 1723, when the King of Sweden obtained a flock, and 1800 most of the countries in western Europe obtained Merinos. Later these populations provided foundation stocks for the Merino breeds of the United States and Australasia.

The most important exports for this future development were to Saxony in 1765, to France from 1750, to England in 1787 and to South Africa two years later. In Saxony, the Elector Xavier, a relative of the King of Spain, selected for even finer wool, and for a period until Australian production came to dominate world trade from about 1850, Germany was the greatest producer and exporter of fine wool.

The Saxony Merino was later exported to

Australia to found the very fine-wooled Tasmanian and Victorian Merino strains. In France, Louis XVI established the Rambouillet flock from which the American Rambouillet breed was developed between 1840 and 1860. In England, George III kept his "Spanish flock" at Kew. There was great interest in developing fine wool production in Britain to supply new wool manufacturers in Yorkshire, but the Merino did not thrive. A few Merinos from Kew were taken to Australia and together with some from the Cape area of South Africa they formed the nucleus for the Australian Merino.

The Merino breeds are distinguished from the other breeds by uniform fineness of the wool they produce which results from both the primary and secondary wool follicles producing fine fibers. The ratio between secondary and primary follicles is about 10:1 in the Spanish Merino and about 20:1 in the Australian Merino—far higher than in other breeds, where the maximum ratio is about 6:1. This gives a very high density of follicles in the skin. The fleeces shorn from Merinos contain a large amount of grease and the yield of clean wool may be as low as 35–50 percent in European Merinos.

◀▲▶ **Longwool, down and carpet wool breeds** of sheep. (1) Dalesbred ewe and (2) Scottish Blackface, hardy mountain and hill breeds with exceptionally springy wool used in carpet and tweed manufacture. (3) Border Leicester ram. (4) Wensleydale ewe. (5) Suffolk ewe. (6) Corriedale ewe, a New Zealand cross between longwools, especially Lincolns, and Merinos; the most common breed in Chile and Argentina, it produces better lamb carcasses than either parent breed. (7) German Blackface ram, a good mutton breed derived from Hampshire Down and Oxford Down.

All the Merino breeds show great similarity. Ewes are generally polled (naturally hornless) but rams, with very few exceptions, have strong, spirally coiled horns with deep grooves. Fleeces are white, although there are black Merinos in Portugal and Tunisia. Wool grows on the top of the head but the face below the eyes is usually bare revealing pronounced skin wrinkles.

In western Europe the value of wool has declined relative to lamb carcasses, and the numbers of Merino sheep have decreased, particularly in Spain and Italy. The remaining Merinos, with the exception of some in Germany and France, are not well suited to lamb production as the breeds are not prolific and do not have good growth rates or good carcass conformation. In the centrally planned economies of eastern Europe, production of fine wool is still increasing and new Merino breeds are still being bred.

Longwool Sheep

In medieval times white-faced sheep with relatively long, medium-fine wool, derived from sheep imported by the Romans, were common in the Midland and Cotswold area of England. By the 18th century, these longwool breeds were found in Kent, the Cotswolds, Leicester, Lincoln and Yorkshire. With the increased demands for food from the growing industrial population, improvements in agriculture were occurring rapidly and the first systematic selection and improvement of sheep for meat production was made in Longwools.

Improvement of the Leicester breed was started by Joseph Allon and others. The major improvements were made by Robert Bakewell (1725–95) of Dishley Grange, Leicestershire. By 1770, by crossing with the Ryeland, selecting and inbreeding, Bakewell had developed a type that was called the New or Dishley Leicester, with much improved meat production and carcass conformation. The Dishley Leicester was widely used in Britain to improve many breeds, but particularly the other longwool breeds. The new longwool breeds have had an impact on sheep breeding throughout the world, second only to that of the Merino.

The longwool breeds produce large fleeces (12–17.5lb—5.58kg) of medium diameter white wool—0.0011–0.0016in (28–40 micrometers) which is usually lustrous. Most breeds are large-framed and heavy. Both sexes are polled.

On a number of occasions in different parts of the world the improved longwools have been crossed with Merinos to produce new breeds that combine good growth and carcass characteristics with the production of large fleeces of medium diameter wool. This type of crossing was first made in France, then in New Zealand, Australia and the USA and is still continuing in the USSR and eastern Europe.

Down Sheep

In the 19th century the shortwool breeds of England were selected and crossed to produce another group of breeds specialized for meat production. The Down breeds as they came to be known, are large-sized breeds that transfer rapid growth and good carcass conformation to their progeny. All are now polled in both sexes, although the local breeds they originate from were generally horned. In Britain the majority of slaughter lambs are sired by one of these breeds. Although they have been exported to many countries, including New Zealand, Australia, the United States and many in Europe, they do not there have the same importance.

Carpet Wool Sheep

Coarse wool, usually with a medulla (hollow core) and an average fiber diameter of more than about 0.0014in (36 micrometers) is termed carpet wool. The carpet wool breeds derive from sheep selected in the first stages of domestication with a coarse outer coat and a finer undercoat. The number of secondary follicles, which produce finer fibers, is low and the ratio of secondary to primary follicles is about 3:1. The fibers may reach 10–12in (25–30cm) in length. The long, coarse fleece sheds rain and snow. Consequently these sheep are well adapted to the harsh environments in which they are found in Asia, northern Europe and in mountain and desert regions at lower latitudes. They have not been exported in large numbers to the New World and Australasia. In New Zealand, Drysdale sheep fill this role.

Although the main use of this coarse wool is in carpet making, where no synthetic fiber has been produced to match its springiness, carpet wool is also used for filling mattresses and making coarse cloths, including tweeds. Making rugs and pile carpets is an extremely ancient craft started by the nomads of Central Asia. Remains exist of Scythian carpets dating from about 1000BC. Carthaginian carpets were traded in Greece in the 5th century BC.

Carpet wool is a secondary product from these breeds. The main product is meat or milk. Many can still be considered as triple-purpose breeds but there is an increasing tendency to select either for improved milk

◀ **Feeding on winter hay,** Olchon Valley, Herefordshire, England. Although kept on good pasture, these longwool crossbred sheep require a winter feed supplement. Longwool breeds are kept primarily for meat production, but the wool is useful for blankets and coats.

▼ **Karakuls in Namibia.** Originating from Turkestan this breed is renowned for lamb's fur. Known as Astrakhan, or Persian lamb, the pelts have tight, lustrous curls of wool, usually black. They are taken when lambs are two days old. The ewes are also useful dairy animals, and their fat tail is a delicacy. Over 12 million are kept in the Soviet Union, 4 million in Namibia.

yield or better meat production. They have few characteristics in common other than their fleece and so differ widely in almost every other respect. The harsh environments result in relatively low weights.

Fat-tail and Fat-rump Sheep

The origin of fat-tail and fat-rump sheep is in Asia and they are still mainly found there, although in ancient times they spread to North Africa. There are many pictures and bas-reliefs of them in Mesopotamia and Egypt dating from about 3000BC. In the last century a few sheep from Somalia were taken to South Africa, where they were erroneously called Blackhead Persian. This breed is now also found in the Caribbean and South America. There are small numbers of semi-fat-tailed sheep in Italy and Greece.

In these sheep most of the fat reserves are found in the tail or rump, instead of in the body cavity and under the skin. Although it has been suggested that this is an adaptation to hot climates, many of these breeds are found in Central Asia where there are extremes of both heat and cold. Man must have selected for this character, as the rams often have greater difficulty in mating fat-tailed ewes.

The size of the tail varies widely with the availability and quality of feed through the year and can be as much as 15–20 percent of the carcass weight. The tail is a delicacy and is sold separately from the carcass. Recent analyses have found that the fat has a higher proportion of unsaturated fatty acids than the other fat reserves of sheep.

Hair Sheep

Hair sheep have a small fleece of coarse fibers with an average diameter of more than 0.0016in (40 micrometers). The undercoat of wool is very sparse and often nonexistent in tropical areas. This distinguishes them from carpet wool breeds (see above). Hair sheep spread, at an early date, from the center of domestication of sheep in southwest Asia. Long-legged, thin-tailed hair sheep with almost horizontal, spiral horns, very similar to the Uda breed in Nigeria today, were depicted in Egyptian paintings around 3000BC. They are now common in Africa, where there are 90 million south of the Sahara, and in India where there are 10 million in the southern peninsular region. In Caribbean islands and the Central American countries bordering the Caribbean, there are 2 million derived from West African sheep taken there at the height of the slave trade in the 16th and 17th cen-

turies. They are still amost identical in appearance to sheep found today in their ancestral West Africa.

These sheep have a primitive fleece type, which is undoubtedly an advantage in tropical and subtropical areas. The hair is generally colored and often spotted, with a wide range of colors including black, brown, red and tan, but the breeds from the Sahel zone of Africa are white or brown and white. Most are small and slow growing but this may be largely due to the harsh environmental conditions in which they are kept. The breeds kept by nomads in the Sahel have longer legs and are heavier than those from the forest zone which have been called dwarf types, because of their small mature height. All the tropical hair sheep have long, thin tails reaching to the hocks. Males usually have horns and may also have a mane or ruff of long, coarse hair.

Hair sheep are kept for meat production. The adults survive well in tropical conditions but losses of young lambs are often high because of poor nutrition and parasitic diseases. Ewes are sexually active throughout the year so the interval between lambings is often less than a year, which can result in annual lambing percentages (the numbers of lambs weaned per 100 ewes put with ram) above the level of prolificacy (the average number of lambs per ewe per lambing). The meat is often consumed by the family, but the sale of live animals may be an important source of extra income. Hides are a by-product but the hair is generally so sparse and coarse that it is not shorn. Very occasionally the ewes are milked.

Primitive Sheep of Northern Europe

Small populations of ancient tribes of sheep with colored and often hairy fleeces and short tails still survive in northern Europe. They give some idea of how sheep kept in the Bronze and Iron Ages looked.

In the St Kilda Islands on the extreme edge of the Hebrides in Scotland, the Soay sheep appear to be survivors of the Bronze Age. They have been brought to the mainland and are now common in zoos, wildlife parks and hobby flocks all over Britain.

The Iron Age sheep survive in a group of breeds with short tails and fleeces similar to the Soay but in a range of colors, including black, gray, brown, and white. This group of breeds include the Romanov in the USSR, the Finnish Landrace, the Goth in Sweden, the Spaelsau in Norway, the Faroe and the Icelandic. In Britain these breeds are associated with the areas settled by the Norse invaders.   TTT

SHEEP BREEDS

Merino Sheep

Spanish Merino

Farmed and originated: Estremadura, Spain. About 25% of Spanish sheep. Fleece: di 6–22μm; sl 1.5–3.9in; wp: 4.4–7.7lb. Prolificacy: 1.1–1.15. wt: ram 88–154lb, ewe 66–121lb. ht: ram 21–27in, ewe 17–21in. Features: rams with slight skin folds. Few flocks now migratory.

Rambouillet

Farmed and originated: France. Sole flock at *Bergerie nationale*. Fleece: di 20μm; sl 2–2.7in; wp: 4.4–7.7lb, Prolificacy: 1.4. wt: ram 154–200lb, ewe 99–132lb. ht: ram 27–29in, ewe 23–26in. Used to improve other Merino breeds, especially the American Rambouillet.

Merino Précoce

Farmed and originated: France. Fleece: di: 18–22μm; wp: 8.8lb–11lb. Prolificacy: 1.2–1.5. wt: ram 220–264lb, ewe 143–176lb. ht: ram 27–31in, ewe 25–27in. Features: bred for early development and carcass conformation. Used mainly in crossbreeding to improve meat characteristics of other Merino breeds.

Merino d'Arles

Farmed and originated: France. Developed mid 18th century from crossbreeding Spanish Merino with sheep of Arles region. Fleece: di: 20–25μm; wp: 6.6lb. Prolificacy: 1.15. wt: ram 154–198lb, ewe 99–121lb. ht: ram 23–31in, ewe 21–27in. Still kept in a transhumant system, wintering on lowlands in Rhone Delta, grazing alpine pastures in summer.

Gentile di Puglia

Farmed and originated: Southern Italy. Developed from 15th to 18th centuries by crossing Spanish Merino with local fine-wooled sheep. Fleece: di: 18–22μm; sl: 3.9in; wp: 6.6lb. Prolificacy: 1–1.5. wt: ram 132–154lb, ewe 83–88lb. ht: ram 25–27in, ewe 22–24in. Milk production. 66–77lb per annum. Once considered a triple-purpose breed, now of reduced importance as not well suited to milk or meat production and income from wool very low.

Deutsches Merinolandschaf
Deutsches Merinolandschaf or Landrace or Wurtemberg Merino or Race de l'Est à Laine Merino.

Farmed: Southern Germany and Alsace region of France. 40% of West German sheep. Bred by Duke of Wurtemberg from Spanish Merinos imported from 1786. Fleece: di: 21–24μm; sl: 2.7–3.5cm; wp: 8.8–10lb. Prolificacy: 1.2–1.5. wt: ram 242–286lb, ewe 143–165lb. Features: good large carcass with big eye muscle area.

Mutton Merino
Mutton Merino or Deutsches Merinofleischschaf.

Farmed and originated: Hanover—Braunschweig region; 5% of West German sheep. Fleece: di: 21–24μm; sl: 2.7–3in; wp: 8.8–11lb. Prolificacy: 1.4–1.6. wt: ram 264–308lb, ewe 154–176lb. Features: combines good carcass characteristics with fine wool production. Ram used for crossing for meat production in hot climates.

Australian Merino

Farmed and originated: Australia. Three strains: fine-wooled small Merinos derived from original imported Spanish and Saxony Merinos; medium-wooled Peppin Merinos derived by crossing small strain with Rambouillet; larger "strong-wool" strain developed in S Australia from Peppin. Fleece: di: fine—less than 20μm, medium—21–23μm, strong—more than 25μm; sl: 2.7–4.7in; wp: 7.7–13lb. Prolificacy: 0.6–1 wt: ewe 88–132lb. Fine-wooled type kept in the high rainfall areas of Tasmania and the tablelands of New South Wales and Victoria. In the higher areas (Southern Australia, Western Australia, western New South Wales and Southwestern Queensland) the strong-wooled type predominate.

American Rambouillet

Farmed and originated: USA, from French and German Merinos imported 1882–1920. The most numerous breed in the USA in the last 40 years. Prolificacy: 1.2. wt: ram 220–275lb, ewe 143–200lb. Features: strong selection against neck wrinkles to make shearing easier.

Caucasian Merino

Farmed and originated: Caucasus. Largest of four Merino breeds in Caucasus, one of at least 10 in USSR developed from strains imported from France and Saxony from 1802. Fleece: di: about 22μm; sl: 2.3–3in. wp: 6.6lb. Prolificacy: 1.5. wt: ewe 154lb.

Longwool Sheep

Leicester

Farmed and originated: England, mainly in East Yorkshire. Few flocks remain of the breed that was improved by Robert Bakewell. Fleece: di: 34–39μm; sl: 7.8–10in; wp: 11–13lb. Prolificacy: 1.56. wt: ram 286lb, ewe 209lb.

Border Leicester

Farmed: England. Originated: Northumberland late 1800s, from crossing Cheviot with Dishley Leicester. Fleece: di: 31–34μm; sl: 6–10in wp: 6.6–10lb. Prolificacy: 1.9–2. wt: ram 220–275lb, ewe 176–220lb. Rams cross with hardy hill ewes, resulting ewes crossed with Down rams to produce slaughter lambs. A similar system is used in Australia to produce crossbreeds from Merino ewes for lamb production.

Blue-faced Leicester

Farmed and originated: England. Developed late 1800s by crossing Border Leicester with Wensleydale. Fleece: black on head; di: 30–33μm; sl: 5–6in; wp: 6.6lb. Prolificacy: 2. wt: ewe 187lb. Slightly more prolific than Border Leicester; used in same way.

Wensleydale

Farmed and originated: England. Derived from an outstanding ram produced in 1838 at Appleton, Yorkshire, by crossing Leicester ram with Teeswater ewe. Fleece: head blue-black; di: 44–48μm; sl: 8–12in; wp: 13lb. Prolificacy: 1.8. wt: ram 308lb, ewe 231lb. Features: exceptionally large; produces progeny entirely without kemp when crossed with ewes that have a proportion of kemp fibres in their fleeces.

Lincoln

Originated: England. Improved with Dishley Leicester. Fleece: di: 36–40μm; sl: 12–16in; wp: 11–14lb. Prolificacy: 1.4. wt: ewe 200lb. Widely exported for crossing with Merinos. Few flocks remain in England.

Kent
Kent or Romney Marsh

Recorded on Kent coast of England for 700 years. Fleece: di: 28–34μm; sl: 6–8in; wp: 8.8–10lb. Prolificacy: 6–8in; wp: 8.8–10lb. Prolificacy: 1.4. wt: ewe 154lb. Features: survives well in harsh conditions at high stocking rates with a minimum of husbandry.

New Zealand Romney

Farmed and originated: New Zealand, from Kent introduced 1853; 90% of country's sheep. Fleece: di: 34μm; sl: 0.7in; wp: 10lb. Prolificacy: 1.2. wt: ewe 110–121lb. Kept in wide variety of environments.

Texel

Farmed: Europe. Originated: Island of Texel, Netherlands, with influence of English breeds, particularly longwool in late 1800s. Fleece: white with black nose tip; di: 29–30μm; sl: 6–8in; wp: 10lb. Prolificacy: 1.75. wt: ram 264–286lb, ewe 165–176lb. ht: ram 29–32in, ewe 26–28in. Features: lambs with good carcass conformation and slightly higher lean to bone ratio than Down breeds.

Charmoise

Farmed and originated: Vienne region of France. Developed from Kent rams crossed with local ewes 1850s in Loire-et-Cher. Fleece: di: 26μm; wp: 4.4lb. Prolificacy: 1.2. wt: ram 165–200lb, ewe 88–110lb. ht: ram 27in, ewe 25in. Features: carcass of good conformation, especially leg.

Bleue du Maine

Farmed and originated: Maine region of France. Developed with influence of English breeds, particularly the Wensleydale. Recognized as breed 1927. Fleece: di: 28–30μm; wp: 8.8–10lb. Prolificacy: 1.9. wt: ram 242–264lb, ewe 176–200lb. Features: woolless blue head; high milk yield resulting in very rapid growth of lambs.

Ile de France
Ile de France or Dishley Merino

Originated: Paris Basin in France. Created in 1832 by crossing Dishley Leicester with Rambouillet. Fleece: di: 23–29μm; sl: 4in; wp: 8.8–10lb, Prolificacy: 1.3–1.5. wt: ram 242–275lb, ewe 143–176lb. ht: ram 26–29in, ewe 26–27in. Features: long breeding season, ewes lamb September—May with most in late fall. Used in the Mediterranean countries as crossing sire for lamb production.

Berrichonne du Cher

Farmed: France. Originated: Berry region of France from local strains with Spanish Merino improvement 1700s, Dishley Leicester and other English breeds early 1800s. Fleece: di: 25–29μm; wp: 6.6lb. Prolificacy: 1.3–1.5. wt: ram 200–242lb, ewe 154–176. ht: ram 26–28in, ewe 25–27in. Features similar to Ile de

France but hardier; shorter breeding season lasting October–March. Rams widely used in France in crossing for lamb production.

Corriedale

Farmed: New Zealand, Australia, Chile, Argentina, Uruguay, USSR and E Europe. Originated: New Zealand from crosses of Merino with Longwool breeds, particularly the Lincoln. Fleece: Di: 27μm; sl: 6in. wt: ewe 132–143lb.

Polworth

Farmed: Australia, South America. Originated: Victoria, Australia from Lincoln-Merino crosses backcrossed to Merino. Fleece: Di: 25–27μm; sl: 4.7in: wp: 12lb. wt: ewe 110lb.

Down Sheep

Southdown

Farmed: Britain, France, New Zealand. Originated: Sussex, England, in 1780s from a local breed found on the hills (downs) with influence of Berkshire Nott. French variety from 1855 imports, larger framed, now more numerous than in Britain. Fleece: Di: 24–27μm; sl: 2–3in; wp: 4.4–5.5lb. Prolificacy: 1.47. wt: ram 154–200lb, ewe 110–154lb. Widely used in New Zealand on Romney ewes to produce lambs for export. Large contribution to the Vendéenne and Charollaise breeds in France.

Suffolk

Farmed: worldwide. Originated: England from 1786, by crossing Southdown rams with Norfolk Horn ewes. Fleece: Di: 25–30μm; sl: 3in; wp: 5.5–6.6lb. Prolificacy: 1.7. wt: ewe 187–200lb. Features: black, hornless head without wool. More slaughter lambs in United Kingdom sired by this breed than any other. Larger frame and weight in USA.

Hampshire Down

Farmed: England. Originated: Berkshire, England, from crossing Southdown, Berkshire and Wiltshire Horn. Fleece: Di: 26–30μm; sl: 3–3.5in; wp: 5lb. Prolificacy: 1.4. wt: ram 220lb, ewe 165lb. Features: black nose and ears with white wool extending over top of head to below eyes.

Oxford Down

Farmed: England. Originated: Oxfordshire, England, in mid-1800s by crossing Hampshire Down with Cotswold. Fleece: Di: 26–29μm; sl: 4–5in; wp: 8.8lb. Prolificacy 1.4.

wt: ewe 200lb. Features: stronger framed, bigger boned, heavier than Suffolk.

German Blackface
German Blackface or Deutsches Schwarzköpfiges Fleischschaf

Farmed and originated: Germany. Derived from imports of Hampshire and Oxford Downs in 1800s; 30% of West German sheep. Fleece: Di: 25–29μm; wp: 8.8lb. Prolocacy: 1.2–1.8. wt: ewe 165lb.

Carpet Wool Breeds

Scottish Blackface

Farmed and originated: Scotland, from sheep imported by Danes. Fleece: Di: more than 36μm; sl: 8–12in; wp:. Prolificacy: 1.75 in older ewes in lowlands. Lambing percentages 80–120 in hill conditions. wt: ewe 88–120lb. Features: both sexes horned; head bare and black with white spots.

Welsh Mountain

Farmed and originated: Wales. Fleece: Di: 30–39μm, often some kemp as well; sl: 2–4in; wp: 2.2–3.3lb. Prolificacy: 1.15. wt: ewe 66–88lb. Features: rams with wide-spiralled horns, ewes polled; hardy, tolerate rain well; head white with dark tip to nose.

Sarda
Sarda or Sardinian

Farmed and originated: Italy; 40% of Italian sheep. Fleece: Di: 37μm; sl: 5.5in; wp: 2.4lb. Prolificacy: 1.1. wt: ram 77–154lb, ewe 66–110lb. Features: ears horizontal, ewes polled but rams may have horns or scurs; tail long and thin; easy to milk by hand or machine, with exceptional yield— 286–474lb—providing 60–70% of income from breed.

Churro

Farmed and originated: Portugal, N Spain. Number of strains with wide range in size. Fleece: Di: 35–55μm; sl: 3.5–6in: wp:3.3–4.4lb. Prolificacy: 1–1.5. wt: ram 55–143lb. ewe 39–121lb. Ht: ram 17–31in, ewe 13–25in. Features: white with distinctive black spots around eyes and on nose; rams with spiral horns, ewes polled. Kept for milk and meat production. Milk yield: about 300lb in 120 days. In 1540 taken by the Spanish to North America; origin of the Navajo breed.

Lacha

Farmed and originated: Basque country of Spain. Fleece: Di: 36–41μm; sl: 10in; wp: 3.3–4.4lb. Prolificacy: 1. wt: ram 88–143lb, ewe 77–110lb. Ht: ram 26–27in, ewe 21–26in. Features: white with dark gray, brownish-black or red head. Horns: spiral in ram, small or absent in ewe. Milk yield: 132–176lb. Kept mainly for milk, rams sometimes fight in competitions. An almost identical breed, the Manech, is found in the Basque country of France.

Karagouniko

Farmed and originated: Thessaly, Greece. Typical of Zakei breeds, which include most sheep in Yugoslavia, Albania and Greece, and some in Romania, Czechoslovakia, Hungary, Bulgaria and southern Italy. Fleece: black, white or brown; Di: more than 36μm; sl: 6in; wp: 3lb. Prolificacy: 1. wt: ewe 88lb. Ht: ewe 25in. Features: rams with laterally spiral horns.

Romanov

Farmed and originated: Yarslavl, USSR. Fleece: gray mixture of black hair and fine white fibers; Di average 38μm, range 18–90μm; sl: 1.1–3in; wp: 5lb. Prolificacy: 2.2–2.6: litters of 4–5 common. wt: ewe 99–121lb. Ht: ram 27in, ewe 26in. Features: short, thin tail; ewes polled, rams may have horns; very hardy, Kept for meat; 6 month lambs produce valuable pelts. Imported to France to improve prolificacy. The Finnish Landrace has a similar origin and has similar prolificacy but, with a few exceptions, the fleece is of fine, white wool (Di: average 25μm, range 14–40μm).

Marwari

Farmed and originated: Rajasthan and Gujarat, India. Fleece: white with black head and neck; Di: 37μm; sl: 2.3in; wp: 4lb. Lambing percentage 87. wt: ram 68lb, ewe 55lb. Ht: ram 24in, ewe 23in. Features: medium-length thin tail; small tubular ears; both sexes polled. Used for meat and wool production.

Deccani

Farmed and originated: central peninsular region of India. Fleece: black, black with white or brown spots on head or white with brown spots; Di: 52μm; sl: 3in; wp: 3lb. Lambing percentage 85. wt: ram 86lb. ewe 62lb. Ht: ram 26in, ewe 25in. Features: short thin tail: long drooping ears; bare belly and legs. Kept for meat production.

Fat-tail and Fat-rump Breeds

Ak-Karaman
Ak-Karaman or White Karaman

Farmed and originated: Anatolia; 50% of Turkish sheep. Fleece: white with black or brown markings on eyes, ears and nose; Di: 40μm; sl: 4.7in; wp: 2.6–4lb. Tail: bare on underside and consisting of three parts—a large oval cushion of fat, a smaller heart-shaped middle piece and thin hanging end usually turned sideways. Milk yield: 44–77lb. Horns: absent except in 10% of rams. Prolificacy: 1.0. wt: ewe 99–121lb. Ht: ram 26–28in, ewe 24–26in. Other features; hardy, surviving dry hot summers and very cold winters with poor nutrition; legs and belly bare; ears long and drooping. Triple-purpose breed.

Awassi

Most common breed in Syria, Lebanon, Jordan, Israel, Iraq. Fleece: white, with red-brown head and legs; coarse and wavy; Di: 32–40μm; sl: 6–8in; wp: 3.7lb. Tail: 10lb covered by long wool: broad fat cushion bare on underside extending to hocks, narrow middle part and thin end. Milk yield: 880lb for highly selected ewes in Israel. Horns: long and spiralling in rams, ewes polled. Prolificacy: 1.25. wt: ram 200–264lb, ewe 66–110lb. Ht: ewe 23–27in. Other features: good walker kept by nomads; resistant to heat but not to cold or humidity, especially snow; ears long and drooping. Triple-purpose breed; milk most important product in Israel.

Karakul

Farmed: 12 million in Turkistan, USSR; 4 million in Namibia. Originated: Turkistan. Fleece: black becoming brown then gray with age; carpet wool, Di: 32μm; sl: 6–8in; wp: 6lb. Tail: broad and shield-shaped at top with small, thin lower part. Horns: usually long and spiralling in rams, ewes polled. Prolificacy: 1. wt: ram 110–132lb, ewe 88–110lb. Features: head bare, nose curved. Kept for Astrakhan (Persian lamb) pelts obtained by slaughtering lambs within 2 days of birth; lamb pelts with tight lustrous curls of wool, mostly black but grays and browns selected.

CONTINUED ▶

Tunisian Barbary

Farmed and originated Tunisia, dating from Carthaginian period. Fleece: white with black or brown head and shoulders; DI: 31–36μm, occasionally as fine as 26μm; WP: 4.4–5.5lb. Tail: 6.6–11lb reaching to hocks with fat, U-shaped upper part, small middle piece and terminal part varying in size with condition of animal. Horns; large and spiralling in rams, ewes polled but often with scurs. Prolificacy: 1. WT: ram 121–154lb, ewe 77–110lb. HT: ram 23–31in. Features: long-legged and hardy, well adapted to desert migration; ears long and pendulous. Kept for meat and wool.

Mongolian Fat-tail

Farmed and originated: Mongolia, Inner Mongolia and adjoining regions of China. Fleece: white with black head and shoulders; outer coat 2.2–4.4lb very coarse and kempy with SL 8in; fine undercoat with DI: 15–20μm; can be combed out in early summer. Tail: wide at top with lower part turned up. Milk yield: 55–66lb. Horns: strong and curved in rams, ewes polled. Prolificacy: 1 WT: ram 99–200lb, ewe 77–154lb. HT: ram 26in, ewe 24in. Features: ears pendulous. Milked in Mongolia. Kept for meat and wool only in China. Was source 800–1,000 years ago of several of the fat-tail breeds found in other parts of China, including the Hu.

Hu

Farmed and originated: Chekiang and Kiangsu provinces of China. Fleece: 2.2lb. Tail: wide at top with lower part turned up. Horns: both sexes polled. Prolificacy: averaging more than 2; litters of 4–6 common. Ewes may lamb more than once per year, lambing percentage more than 300. WT: ram 77–132lb, ewe 66–99lb. HT: ram 26in, ewe 25in. Kept in sheds throughout the year in densely populated regions without grazing, fed on vegetable waste, mulberry leaves and grass cut from verges. Dung an important by-product for fertilizing vegetable crops and mulberry trees for silk production. Lambs slaughtered at 2–3 days for white pelts with lustrous, curled wool.

Chios

Chios or Sakiz

Farmed: Greece, Cyprus and Ismir area of Turkey. Originated: Greek island of Chios. Fleece: white with black or brown spots around eyes and on nose; DI: 32μm; SL: 3in; WP: 3.3lb.

Tail semifat, cone shaped, about 10in long 3.5–5in at top. Milk yield 374–440lb. Horns: long and spiralling in rams, ewes generally polled. Prolificacy: 1.8–2 WT: ram 143–176lb, ewe 110lb. HT: ram 32in, ewe 27–30in. Features: long-legged with fine bones. Kept in very small flocks by citrus growers in lowland Chios.

Blackhead Persian

Farmed and originated: South Africa. Developed from 3 sheep imported from Somalia in 1868. Small numbers in Trinidad and Tobago, Venezuela, Colombia, Brazil. Fleece: DI more than 40μm; short, sparse. Tail: small, thin. Fat-rumped. Horns: both sexes polled. Prolificacy: 1.1–1.15. WT: ewe 59–79lb. Features: pronounced dewlap. Crossed in South Africa with Dorset Horn to produce the Dorper breed.

Hair Sheep

Nellore

Farmed and originated: Nellore, Prakasham and Ongole districts of Andhra Pradesh state of India. Fleece: three varieties with little hair except on brisket, withers and breech—**Palla** white with light brown spots, **Jodipi** white with black spots around eyes and lower jaw, **Dora** completely brown. Tail: short—4in—thin. Horns: heavy in rams, ewes polled. Prolificacy. 1. Lambing percentage: about 75. WT: ram 81lb, ewe 66lb. HT: ram 30in, ewe 28in. Features: ears long and drooping.

West African Dwarf

West African Dwarf or Djallonke

Farmed and originated: humid forest zone of West Africa. Fleece: black or black and white spotted. Horns: short and crescent-shaped in rams. Prolificacy: 1–1.2. Lambing percentage up to 140. WT: 44–55lb. HT: 16–18in. Features: tolerant of trypanosomiasis; legs very short; males with well-developed mane. Traditionally roam free scavenging in villages, tethered or penned when threatening crops.

Barbados Blackbelly

Farmed: W Indies, Mexico, Venezuela, USA. Originated: Barbados from W African sheep, possibly from Cameroon, where similar sheep are common, and from a northern European strain. Fleece: light to dark reddish-brown, black belly, head and legs. Prolificacy: 1.85–2. WT: ram 154–200lb, ewe 88–132lb. Features: rams with horns and large black mane.

Primitive Sheep of Europe

Soay

Farmed and originated: St Kilda Islands, Scotland. Fleece: brown, 75% dark, 25% light with white belly, hairy type with outer coat of hairy fibers of fine kemp and small undercoat of very fine wool; average DI 30μm, range 12–144μm; wooly type average DI 25μm, range 12–50μm; SL: 2in; WP: 0.6–1.5lb; molt in May–June. Horns: strong in single whorl in rams; ewes horned or polled. Prolificacy: 1. WT: ram 79lb, ewe 55lb. HT: ram 22in. Features: tail short like the wild types of sheep.

Orkney

Orkney or North Ronaldsay

Farmed and originated: Orkney Islands, Scotland, now largely restricted to the Island of North Ronaldsay. Fleece: white or gray; black and light brown less common: DI: average 22μm, range 8–120μm; SL: 2–6in; WP: 2.2–2.6lb, natural molt in spring. Horns: strong, curving in rams; ewes generally polled. WT: ewe 66–77lb. Kept on seashore by a wall except for a short period around lambing. Diet consists mainly of seaweed.

Herdwick

Farmed and originated: Lake District, England. Fleece: black in lambs, turning gray in older sheep; DI: 28–32μm; SL: 6in; WP: 3.3–4.4lb. Horns: large in rams; ewes polled. WT: ewe 77–88lb. Features: exceptionally hardy and tolerant of rain.

Shetland

Farmed and originated: Shetland Islands, Scotland. Fleece: white, gray, light brown; DI: average 26μm, range 12–60μm; SL: 4–5.5in; WP: 2.2–3.3lb: tendency to molt, although now shorn annually. Horns: large, curved in rams; ewes polled. WT: ewe 66–77lb. Features: fine, soft wool used for high-quality tweed and hand-knitted garments.

▶ ▼ **Primitive, hair and fat-tail sheep. (1)** Soay ram, molting. **(2)** Manx Loghtan ram, a primitive Viking type with naturally short tail and four horns, as here, or two. **(3)** Nellore ewe, Jodipi variety. **(4)** Sudanese, a desert hair sheep found in the African Sahel, with fat deposits down either side of tail when in good condition. **(5)** West African Dwarf ram. **(6)** Algarve Churro ewe, a coarse-wooled Portuguese breed derived from the Churro, mainly used for meat. **(7)** Hu ewe. **(8)** Mongolian fat-tail ram.

6

GOATS

Capra hircus aegagrus
Order: Artiodactyla
Suborder: Ruminantia
Family: Bovidae
One of the seven species of genus *Capra*
Number of breeds: approximately 216

Distribution: worldwide in warm and temperate zone, with 70 percent in Asia and Africa. Feral populations worldwide, with largest in New Zealand. Total world population (farmed and feral) about 445 million.

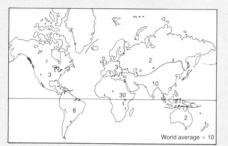

Number of goats, by region, relative to 100 head of population (FAO data 1982).

Size: males, height at withers 16–40in (40–100cm); weight 27.5–275lb (12.5–125kg); females, height at withers 16–40in (40–100cm); weight 22–220lb (10–100kg).

Coat: varies from long (7–8in, 18–20cm) in Angora to short (about ½in, 1cm); straight to finely curled; hair fine in Cashmere breed, but coarse in most others, with finer fur under top fur. Color plain gray or gray combined with white, fawn, brown, red or black. Males often have beards.

Horns: absent (polled), small to large, straight, curved or cork-screw shape; less prominent and less curved in females.
Other features: nose usually either straight or domed; tail 7–10in (18–25cm) long, held erect.

Diet: varied; woody browse preferred.

Metabolism: temperature 101.0–105.5°F (38.4–41.0°C); respiration rate 10–20/min; pulse 70–80/min.

Dental formula: I0/4, C0/0, P3/3, M3/3.

Longevity: 10–12 years.

◄▶ **Goats in profile.** ABOVE The short-haired Anglo-Nubian, with its drooping ears and convex nose, is a polled (hornless) breed suitable for intensive dairying. RIGHT The Angora, with short, horizontal ears, straight nose and long, twisted horns is best known for its long, lustrous wool.

F OR families in some countries goats are the main source of food and clothing. Both economic and cultural pressures encourage this situation. Small peasant families can consume the meat of an entire beast, avoiding the need to develop preservation methods. Goats can be raised by families that either cannot afford to raise cattle or live where the climate is too hot to do so. In some areas religious practices discourage the consumption of pork or beef, but do no forbid the eating of goat meat.

The wild goat, as an identifiable species, probably existed 7 million years ago. The earliest evidence of domestication comes from Jericho in Jordan and from Belt Cave on the shore of the Caspian Sea, where remains from levels dating back to 7000BC have been found. Domestication does not seem to have modified man's regard for goats in the West. The goat's agitated character encouraged men to associate the animal with presence of evil spirits. As late as the Middle Ages goats were identified with Satan and with witches—one foot of the devil was often depicted in the form of a goat's hoof. Only in countries where the goat was of considerable economic importance did it achieve a higher status. In Africa, for example, the goat has been considered an important sacrificial animal, or even as a deity.

The domestic goat of today almost certainly originated from the wild goat, bezoar or aegagrus (*Capra aegagrus*) which is still found on Crete and on other Greek islands and in Turkey, the Middle East, Iran, Pakistan and parts of Afghanistan. It was the first ruminant and one of the first animals of any type to be domesticated. It can be difficult to distinguish between some breeds of domestic goats and some breeds of domestic sheep. Goats can be recognized from their beards and from the presence of scent glands beneath their tails. Domestic goats also tend to carry their tails erect whereas sheep normally let theirs hang down. A more important biological difference exists, however. Domestic goats have 60 chromosomes per cell, domestic sheep 54. It is impossssible to produce viable hybrids. The domestic goat has been crossed, though, with the intermediate species of wild sheep, the Barbary (*Ammontragus*), which has 58 chromosomes.

In temperate zones goats are seasonal breeders. Estrus (or heat) usually begins in September and lasts for five months (ie September–January inc.); maximum length is seven months (ie August–February inc.). If mating does not occur immediately the

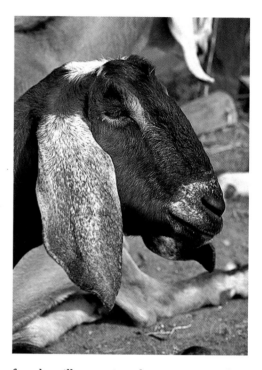

female will come into heat every 21 days during the season. Heat lasts for 24–36 hours. In tropical countries goats are not seasonal breeders. It is common for a female to have three or four kiddings within two years. Gestation period is 150 days. Litters average two (though triplets are common in some breeds).

The characteristics that set goats apart from most other farmed animals are their natural friendliness and inquisitiveness. Their friendliness makes them companionable animals to a point where throughout Europe and North America many are kept more as pets than a source of food or income. Their inquisitiveness however, if sometimes impressive, is generally a nuisance in farming, where they will undo gates, destroy fences, eat valuable documents, and generally create havoc if they find their way into areas reserved for man.

Their natural surefootedness, which in a natural environment makes it possible for them to reach seemingly inaccessible food plants, gives them access to areas where they may cause considerable damage to vegetation.

Goats have a voracious appetite for woody browse material and can climb into trees in search of forage. They thus have the ability to attack environments at their weak points, by destroying trees, which can cause the death of plant life, the death of animal life, and the emergence of desert. Whilst goats are, undeniably, destructive to trees and shrubs, serious ecological damage blamed on goats can often be attributed to mismanagement of goat populations, or to

the fact that they are often the only animals left in an environment already spoiled by mismanaged cattle or other stock. Indeed it can be argued that goats, by controlling brush, improve grassland rather than spoil it.

Throughout the world, where man has explored and settled, domestic goats have escaped and have adapted to new surroundings. Some of today's feral populations may be centuries old. The largest are those in Australia and New Zealand. The latter, originating from animals left by Captain Cook, now number several thousand. Surprisingly, feral populations live even in countries as densely populated as the United Kingdom. Quite large numbers of goats can be found in the Snowdonia region of Wales and in the highlands and islands of Scotland.

There are now approximately 445 million domestic goats in the world, representing 16 percent of all ruminants. Seventy-three percent of these goats are to be found in the developing countries. In many areas goats have been introduced only in recent times while in others they have existed for long enough to be regarded as truly indigenous with many local breeds recognized. Areas with significant populations of introduced goats are North America, most of Europe, Australia and New Zealand.

Large populations of indigenous goats are to be found in South America, Africa, the Middle East and most of Asia, including the USSR and the Far East. The largest populations of goats tend to be in the drier countries. This is not so much because the climate particularly suits goats but because the sparse vegetation offers a meager living for goats alone.

Goats are ruminants, which means that their gut system has evolved to digest efficiently cellular plant material by means of several "stomachs," the first and largest being the rumen. Goats are also opportunist browsers; given a free choice they feed on a wider variety of plants than other domesticated species, carefully selecting the plant or parts of the plant they prefer. They have the hardest mouth of any ruminant and show this in their preference for woody plants. Their tough, almost prehensile, lips can even cope with the needle-like spines of the thorny acacia. These features together give goats the ability to digest large quantities of poor-quality forage, which offers great potential for the intensive farming of goats where high levels of production may be achieved from vegetable matter that is considered as waste by other livestock farmers.

▲▶ **Tree-climbers ancient and modern.** ABOVE Moroccan goats in the Atlas Mountains demonstrate balance, agility and a determination to exploit even the roughest forage to the full. RIGHT In a figurine from the royal cemetery in the ancient city of Ur, a goat reaches up to the fruits of the Tree of Life, symbolizing the Mesopotamian King's special relation with Ishtar, goddess of fertility. The figurine dates from 2500–2400BC.

◀ **On an expedition to Everest,** a flock of long-haired goats in Nepal stands ready to carry packs of supplies into the high country. Surefootedness and their ability to live off the land have made goats useful as beasts of burden in many regions not served by roads.

The Role of Goats Today

Each year the domestic goats of the world produce approximately 7,236,000 tons of milk and 1,892,000 tons of meat. In recent years the proportion of milk to meat has decreased, following an increase in the numbers of goats kept principally for meat in the Third World and a simultaneous reduction in the number of goats in Europe and North America. Although in the West and in other industrialized areas of the world the goat is kept chiefly as a dairy animal, on a global scale it must be regarded as a general-purpose animal providing milk, meat, fiber and skins.

Milk. Goats are kept primarily as dairy animals in the United States and in most of Europe, in Australia and in New Zealand. In these countries modern intensive rearing methods are used and farms can now be seen where up to 1,000 goats may be kept and milked in an ultra-modern milking parlor. Such conditions produce a high level of production, particularly of milk. A desirable average output is 260 US gallons (1,000 litres) of milk per goat per year. The milk is composed of 71.8–75.6 percent water, 13.2 percent solids, 3.5–4.5 percent fats, 3–4.6 percent protein, 4–5 percent lactose, and 0.7–0.9 percent ash.

A scientific approach to the breeding of goats for milking began towards the end of the 19th century when goats carried on ships from the East to provide fresh milk for passengers were bought by interested goatkeepers in London. These were crossed with local breeds to produce such crosses as "English and Indian" and "English and Abyssinian," which were later crossed with four male goats, a Jamnapari and two other goats from the Chittal district of India and a Zairabi from Nubia. These males sired the first goats to be registered as "Anglo-Nubians."

The breeds best suited for intensive dairying enterprises are Anglo-Nubians and Saanens. The Saanens and similar breeds from Switzerland are the most productive dairy goats in the world and have often been bred to produce types suited to particular countries, for example British and French Saanens. The highest milk yield ever achieved by a goat came from a Saanen in Australia, Osory Snow Goose, which produced 7,713lb (3,506kg) of milk in 365 days. High yields can only be obtained where feed is of sufficient quality and quantity. It is unlikely that yields above 11lb (5kg) per day (or 4,015lb, 1,825kg per year) could be achieved in the major goat farming countries even if a highly developed breed was used. Certain traits simply cannot be

World Importance of Goats

World Goat Numbers and Products (1979)

	Number (millions)	Goat meat (thousand tons)	Goat milk (thousand tons)
World	445.9	1,894	7,237
Africa	144.7	505	1,365
N and C America	12.0	28	308
S America	18.4	64	134
Asia	253.6	1,169	3,476
Europe	1.6	87	1,554
Australasia	0.15	1	not recorded
USSR	5.5	40	400

Angora Goat Population and Mohair Production (1979)

Country	Number (millions)	Total production (thousand tons)
Republic of South Africa	1.4	6.1
Texas (USA)	1.2	4.1
Turkey	2.0	4.5
Argentina	1.0	1.0
Lesotho	0.8	0.6
World	6.4	16.3

MILK BREEDS

Anglo-Nubian

UK, derived from the crossing of goats
of Anglo-Indian stock with
Jamnapari, other goats of Indian
origin, and with a Zairabi from Nubia
(Egypt/Sudan).
Coat: short, various colors. Features:
horned or polled; ears long and
drooping, nose convex, wt: female
132–176lb, male 176–264lb. ht: female
27–33in, male 35–39in. Milk yield per
annum (p.a.) 1,322–1,763lb.

Balkan

Albania, Bulgaria, Greece,
Yugoslavia.
Coat: long, in any combination of
black, brown, red, and white.
Features: horned, often curved or
"corkscrew"; ears horizontal or
drooping; nose straight, wt: female
66–110lb, male 88–143lb. ht: female
25in, male 27in. Milk yield p.a.
220–352lb.

Barbari

E Punjab (India) and Pakistan. Coat:
short, white, with red or brown spots.
Features: horned; ears erect; nose
straight. wt: 44–55lb. ht: 25–29in.
Milk yield p.a. 330lb. A prolific
breeder, often producing twins or
triplets.

Beetal

Punjab region (India and Pakistan).
Coat: short, red or tan, with black,
gray, or white spots. Features:
horned; ears long, pendulous; nose
convex. wt: 88–132lb. ht: 25–27in. Milk
yield p.a. 440–660lb.

British Alpine

United Kingdom.
Coat: short, black. Features: face
white; legs similar to the
Toggenburg; horned or polled; ears
erect and pointing forward; nose
straight. wt: female 99–121lb, male
143–176lb. Milk yield p.a. 1,543–1,763lb.

Carpathian

S Poland, Romania.
Coat: long, usually gray, but may be
a combination of black, white and
red. Features: horned; ears erect;
nose straight. wt: 77–99lb. ht: 23–27in.
Milk yield p.a. 550–617lb.

French Alpine

France, particularly the Loire Valley;
a similar breed from the same stock
occurs in Italy.
Coat: short, in various colors, but
brown black markings most common.
Features: horned or polled; ears erect;
nose straight. wt: female 77–88lb, male
99–110lb. ht: female 25in, male 27in.
Milk yield p.a. 660lb.

Garganica

Gargano Peninsula, E Italy.
Coat: long, dark chestnut. Features:
horned; ears erect; nose straight, wt:
female 88lb, male 132lb. ht: female 25in,
male 27–29in. Milk yield p.a. 880lb.

Spanish

Spain (including numerous regional
varieties).
Coat: short, black, chestnut or brown.
Features: horned or polled; ears erect;
nose straight. wt: female 77–88lb, male
99–110lb, ht: female 25in, male 27in.
Milk yield p.a. 660lb.

Girgenta

Agrigento Province, Sicily, Italy.
Coat: long, white with brown spots.
Features: horned, screw shaped; ears
short; nose straight, wt: female 88lb,
male 132lb. ht: female 25–27in, male
27–29in. Milk yield p.a. 330lb.

Golden Guernsey

Farmed: Channel Islands (UK).
Probably originates from the crossing
of Mediterranean and Swiss breeds.
Similar goats can be found on Malta.
Coat: long, reddish brown. Features:
horned or polled; ears erect and
pointing forward; nose straight.
wt: female 88–110lb, male 143–176lb.
ht: 25–29in, male 29–33in. Milk yield
p.a. 880–1,100lb.

Jamnapari

Jamnapari or Etawah
India and SE Asia.
Coat: short, white, with black or
brown patches. Features: horned;
ears very long, pendulous; nose
convex, wt: 110–176lb. ht: female
39–42in, male 43–50in. Milk yield p.a.
550–880lb.

La Mancha

USA.
Coat: short, mixed colors. Features:
polled; ears vestigial; nose straight.
wt: female 32–165lb, male 176–220lb,
ht: female 29–31in, male 31–37in. Milk
yield p.a. 1,760lb.

Malabari

N Malabar, Tellicherry, Mangalore
(W India).
Coat: short, light in color mixed with
black and brown. Features: horned or
polled; ears pendulous; nose convex.
wt: 88–132lb. ht: 25–27in. Milk yield p.a.
440lb. A prolific breeder, often producing
twins or triplets.

Malaguenà

S Spain.
Coat: short, fawn to chestnut red.
Features: horned or polled; ears long
and horizontal; nose straight.
wt: female 99–121lb, male 154–176lb.
ht: female 27–31in, male 29–33in. Milk
yield p.a. 880lb.

Mamber

Mamber or Syrian Mountain
Near and Middle East; similar
varieties occur in Cyprus, Egypt and
Libya.
Coat: long and black. Features:
horned, twisted, up to 12in in length in
male; ears long and pendulous; nose
slightly convex, wt: 88–154lb ht: female
28in, male 33in. Milk yield p.a. 330lb.

Murcia-Granada

S Spain.
Coat: short, black or mahogany
brown. Features: horned or polled;
ears small and forward pointing; nose
straight. wt: female 110lb, male 154lb.
ht: female 29in, male 31in. Milk yield p.a.
1,102lb. Normally produces twins.

Nordic

Norway, Sweden and Finland.
Coat: long, usually white but can be
blue-gray and brown. Features:
horned or polled; ears erect and
pointing forward; nose straight. wt:
female 99–121lb, male 154–187lb.
ht: female 27–31in, male 29–33in.
Milk yield p.a. 1,102–1,322lb.

Saanen

Farmed: improved types all over the
world. Originated in the Saane Valley
in W Switzerland.
Coat: short, white; skin sometimes
black spotted. Features: horned or
polled; ears erect and pointing
forward; nose straight. wt: female
132–176, male 176–264lb. ht: female
29–33in, male 31–39in. Milk yield p.a.
1,763–2,200lb; record 7,729lb.

Sudanese Nubian

Sudan and NE Africa.
Coat: long, black or brown. Features:
horned or polled; ears long and
pendulous: nose convex, wt: 881–
154lb. ʜt: 27–29in. Milk yield p.a. 880lb.
The only African breed kept primarily for
milk. Other breeds of the type include the
Zairibi, Damascus, Shami, Kile, Mzabite
and Mishri.

Toggenburg

Farmed: improved types worldwide,
eg British Toggenburg. Originated:
N Switzerland. Coat: fine, long, fawn
to mousy brown with white markings
on head and legs. Features: horned or
polled; ears erect and pointing
forward; nose straight. wt: female
99–110lb, male 143–176lb. ʜt: female
27–31in, male 29–33in. Milk yield p.a.
1,543–1,760lb.

The following breeds, kept primarily
for meat, also produce useful
quantities of milk: Criollo, Dara Din
Panah, Ma T'ou, Small East African.

▶ ▲ **Breeds of goat kept for
milk production.** (1) Murcia-
Granada. (2) British Alpine.
(3) Saanen. (4) Anglo-Nubian.
(5) Toggenburg. (6) Nordic.
(7) Mamber. (8) Balkan.
(9) Jamnapari. (10) Girgenta.

totally eliminated: the Anglo-Nubian, though well suited for dairying, is difficult to handle. Some breeds, for example the Toggenburg and British Alpine, only produce good yields when farmed by a free-range grazing system.

The benefit of goat's milk to the consumer has been recognized since antiquity. According to Greek mythology, Zeus, the king of heaven, was nourished with the milk of a goat. The main value of goat's milk to health is as an alternative to cow's milk when this causes digestive disorders. In countries where infants are reared on cow's milk products there is an increasing tendency towards sensitivity or allergy to cow's milk. In the United Kingdom it is estimated that 5 percent of all infants are allergic. Approximately 75 percent of these will not show the same allergy to goat's milk. Goat's milk is also easier to digest, probably due to the softer curd it forms in the stomach. This may be attributed to its high proportion of small fat globules. For the same reason goat's milk cream does not rise in the same way as in cow's milk and is therefore more difficult to separate.

Tuberculosis was a major disease in human populations before cow's milk was routinely pasteurized; it is virtually unknown in goats. This, along with the absence of *Brucella abortus*, which occurs in cows and causes undulant fever in man, makes the consumption of raw (unpasteurized) goat's milk perhaps slightly less of a risk than cow's milk.

In spite of these advantages, goat's milk represents only 1.6 percent of the world's milk production. This is because most goats live in countries where they are kept primarily for meat and where the level of feeding is not conducive to high milk yields. An additional factor in these countries is the lack of resources or the knowledge to operate breeding and development programs to improve goat production. In France, on the other hand, the average annual milk yield per goat increased between 1960 and 1977 from 580 to 1,115lb (264kg to 506kg).

Goat's milk has a reputation of having a strong or unpleasant taste. It does rapidly develop "off" or goaty taints if handled without respect for hygiene and without care when cooling. The most characteristic taint is due to the breakdown of fats (lipolysis) by enzymes releasing caproic and caprilic acids. This process can be inhibited by rapid cooling or by pasteurization. When produced under correct conditions goat's milk is to some indistinguishable from cow's milk and certainly pleasant to taste. Breeds

of Swiss and British origin produce milk of the highest butterfat content, with the Anglo-Nubian the highest.

In many countries goat milk is traditionally made into a variety of products. Yogurt was originally a goat-milk product, from the Balkan countries. It is produced by fermenting milk with the bacteria *Lactobacillus bulgaricus* and *Streptococcus thermophilus*. The milk sugar (lactose) is converted to lactic acid which acts as a preservative whilst leaving the protein more or less unchanged. As with many goat products, goat's yogurt sells for a higher price than that made from cow's milk.

Goat cheese is a particular speciality of several Mediterranean countries, with perhaps France producing the greatest quantity and variety. The simplest cheese is lactic cheese, produced by curdling milk for up to 24 hours with lactic acid and a little rennet. The consequent curd is drained, put into a mold, and eaten fresh the next day. Soft cheeses are made in a similar way but are curdled more slowly, drained and ripened. Many have a mold crust, as in the French cheeses Sainte Maure, Valency and Crottin. Fewer blue-mold cheeses are made with goat milk. They are produced by injecting *Penicillium* molds after curdling. The cheeses are ripened in very specific conditions, often in caves. Mold Savoy is an example of this type. Hard cheeses are particularly favored in warm countries, where their prolonged keeping qualities are an advantage. They are made by compressing the curds in special molds after which they are ripened for quite long periods. Examples of hard goat cheese are Chevrotin from France and Queso de Cabra de Malaga from Spain.

Meat. Worldwide the major importance of goats is for meat production. Out of the world population of 445 million, 308 million are farmed mainly for meat in Africa and Asia. The widespread use of goats for meat production has not led to the development of a specific meat breed, though the Boer goat from South Africa has been improved for this purpose in recent years; castrated males may grow to 220lb (100kg) with little supplementary feeding.

Very little goat meat is eaten by Europeans or North Americans, even though they prefer meat that is lean, like that from the goat. Often goats have been the subject of ridicule and criticism, and consumption of any goat product has been considered the obsession of fanatical minority groups. However, there are many countries in the world where goat meat is preferred to

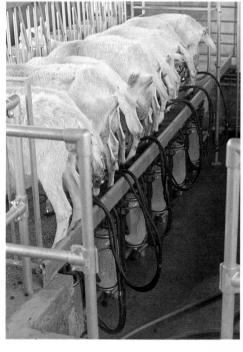

◄▲▶ **Goat husbandry around the world.** LEFT Milk is the most important product of the small goat industries in western countries. Herds are often machine-milked in modern parlors. ABOVE Raised for their meat and their hides, goats share a drinking trough with cattle in Uganda. RIGHT In the mountains of Afghanistan a triple-purpose meat-fiber-and-dairy goat is milked by hand.

mutton or beef. They include Nigeria, the Sudan, India, Malaysia, Fiji and many in the Middle East.

Goat meat has a pleasant flavor, not unlike good mutton or lamb. It is less tender, however, probably because of the changes that occur in muscle fibers when the goat is cold due to their poor covering in carcass fat. The modern trend of not castrating bulls and boars kept for meat would not be suitable for goats because the strong smell of entire males would taint the meat to an unacceptable degree.

Fiber. Evidence suggests that ever since goats were first domesticated their hair and skin have been used by man for clothing. Reference to the weaving of fine goat hair into curtains, for example, can be found in the Old Testament (Exodus 26:7–13).

The breed most highly prized for its fleece is the Angora, the fiber of which is called mohair from the Arabian word *mukhayyar* meaning best fleece. Mohair is fine, lustrous and silky with a staple length of about 8in (20cm). Its main attractions as a commercial product are its high luster, receptivity to dies and great durability. Mohair from young animals is used to produce fashion fabrics; that from mature animals is used for hand knitting yarns, velors, fabric, upholstery

materials, carpets and curtains. The centers of mohair processing and use tend to be away from the producing country. England is the major center for mohair processing.

The Angora goat has been selectively improved to produce more fiber of a better quality so that the average adult fleece will now weigh 8.8–11lb (4–5kg) compared with less than 4.4lb (2kg) at the turn of the century.

The current high price obtainable for the fleece has resulted in the introduction of Angoras to many parts of the world; they are now farmed in Turkey, Texas, Argentina, Lesotho, Australia, New Zealand and South Africa. The latter country produces the greatest quantity of mohair, some 6 thousand tons out of a total world production of 16 thousand tons.

Angoras are very sensitive to climatic stress, particularly when freshly shorn, and therefore need to be kept in an environment with a fairly even temperature. In some countries they have been crossed with other breeds, in an attempt to produce an animal that is tolerant to a wider range of climatic conditions but will still produce high-quality fiber.

In many areas, Angoras are useful in controlling brush on poor-quality range. Unlike many of those of other breeds, Angora males

▲▶ **Breeds of goats kept for meat, hair and other products.** (1) Brazilian. (2) Angora. (3) Kambang Kabjang. (4) Black Bengal. (5) Cashmere. (6) Red Sokoto. (7) Small East African. (8) Pygmy.

Meat Breeds

Abbreviations: wt: weight. ht: height at withers.
Where not specified otherwise, information relates to both females and males. Region refers to where originated and farmed today.

Black Bengal
Bengal (NE India) and Bangladesh. Coat: short, black. Features: horned or polled; ears erect, pointing forward; nose straight. wt: 17–33lb. ht: 15–20in.

Boer
S Africa
Coat: short, various colors (an improved variety is white with red markings). Features: horned or polled; ears long and pendulous; nose convex. wt: 132–200lb. ht: 27–39in.

Brazilian
Brazil, of Portuguese origin. Coat: short, mixed colors (there are four main color types: Canindé black or brown; Morota white; Moxotó white or fawn; Repartida black at front, brown at rear). Features: horned; ears erect; nose straight. wt: 66–77lb. ht: 24–25in.

Criollo
Farmed: S America, especially Mexico, Peru, Venezuela. Originated Spain. Coat: short, mixed colors; the commonest combinations are of black, white, and red. Features: horned; ears erect; nose straight. wt: 66–100lb. ht: 25–29in.

Dara Din Panah
Punjab (India) and Pakistan. Coat: long, usually black, sometimes red. Features: horned, screw shaped; ears extremely long and pendulous; nose slightly convex. wt: 88–100lb ht: 27–32in.

Kambang Kabjang
Burma, Indonesia, Malaysia, Taiwan, Thailand, the Philippines. Coat: short, usually black or black and white. Features: horned; ears erect, pointing forward; nose straight. wt: 44–88lb ht: 19–24in.

Ma T'ou
C China.
Coat: short, white. Features: polled; ears erect, pointing forward; nose straight. wt: 88–132lb. ht: 23–27in.

Small East African
E. Africa.
Coat: short, often light, various colors. Features: horned; ears erect; nose straight. wt: 55–66lb. ht: 23in.

South China Goat
SW China.
Coat: short, black. Features: horned, twisted; ears erect; nose straight. wt: 44–88lb. ht: 19–24in.

West African Dwarf or Pygmy
Guinea, Central W Africa (Mali, Niger, Upper Volta etc), Zaire, Angola, N Namibia.
Coat: short, mixed colors (usually a combination of black or brown with white). Features: horns, short; ears short and erect: nose straight. wt: 44–55lb. ht: 15–20in.

The following breeds, kept primarily for milk, are also important sources of meat: Balkan, Jamnapari (or Etawah), Mamber (or Syrian Mountain).

Fiber or Skin Breeds

Angora
Africa, Australia, C Asia, Lesotho, New Zealand, USA; originated C Asia. Coat: white, hanging in lustrous locks or ringlets. Features: horned, long in male with a twist to right angles; ears horizontal or drooping; nose straight. wt: 66–145lb. ht: 19–27in. Used for fiber.

Cashmere
C Asia.
Coat: long, white, sometimes fawn. brown or gray. Features: horned, long and twisted; ears erect; nose straight. wt: 66–145lb. ht: 23–27in. Used for fiber and skin.

Red Sokoto
Red Sokoto or Maradi
Niger, Sokoto province of Nigeria. Coat: short, mahogany red. Features: horned; ears erect, pointing forward; nose straight. wt: 55–66lb. ht: 23–26in.

The following meat breeds are also prized for their fibers and skins: Black Bengal, Brazilian, Dara Din Panah. Small East African (especially the Mubende and Kigezi varieties).

are not aggressive during the breeding season and do not smell as much or become as filthy. They have a reputation for being poor breeders, a result of poor mothering. When females are kept in open-range conditions they are apt to leave their kids and forget them. In controlled environments this does not happen.

The term Cashmere applies both to a breed of goat, which originates from mountainous regions of Central Asia—not just Kashmir—and to the fine fiber produced from several goat breeds, including the Cashmere. The fiber, also known as pashmina, is the down or underfur that grows under the outer coarse hairs. It is the finest animal fiber used for textiles, having a diameter of less than 0.0006in. The famous ring shawls brought to Europe from India by Marco Polo in the late 13th century were so fine they could be passed through a wedding ring.

For centuries cashmere has been the byproduct of goats mainly kept for meat, including the Cashmere, whose hide is also made into leather and which is even used as a draft animal pulling small carts. Productivity is low—one sweater requires 374–485lb of cashmere fiber, which may be the total annual output from five goats.

Among other breeds that produce cashmere are the Chungwei of China, the Morghose and Raini of Iran, the Kurdi of Iraq, the Vatai from Afghanistan, the Don and Orenburg from Russia and numerous breeds from India and Pakistan including the small breed called pashmina (meaning woolen). The Don goats from the USSR are said to produce more cashmere than any other breed. Yields of 880–3,300lb of fiber per goat per annum are reported.

Skins. Very high quality leather is produced from goat skins such as velor, suede and chamois for clothes and saffian and Morocco for gloves and book binding. It is estimated that about 350,000 fresh goat skins are produced annually throughout the world.

The Red Sokoto, farmed in Nigeria, is one of the few breeds where the hide is the most important product. Their short legs seem to make them unsuitable for long walks to grazing or foraging areas and therefore they are usually maintained under sedentary farming conditions.

Other countries producing high-quality goat skins are China, India, Pakistan, Indonesia and Brazil, and other breeds important for leather productions are the Black Bengal, Brazilian, Small East African Chungwei and Cashmere. AM

PIGS

"Sus domestica" (S. scrofa (Wild boar) and other species and subspecies of the genus Sus contributing to a variety of crossbred forms)

Order: Artiodactyla.
Family: Suidae.
Number of breeds: 87 recognized by breed societies; approximately 225 additional forms with specific appearance or location. Crossbreeding is widespread.

Distribution: worldwide in warm and temperate climates; in heated housing in cold climates. Total world population 780 million, almost half in Asia.

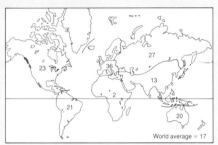

Number of pigs, by region, relative to 100 head of human population (FAO data 1982).

Size: up to 3.3ft (1m); weight 175–440lb (80–200kg); these weights seldom achieved as animals are slaughtered when immature, but liveweights of more than 1,100lb (500kg) recorded; capable of laying down immense quantities of fat.

Coat: straight bristles about 1in (2.5cm) long; some breeds with curly coats; bristle and skin color variable.

Other features: tail curly; prominent snout ending in mobile disk-like nose pierced by nostrils; ears huge pendulous to short pricked.

Diet: varied, a true omnivore.

Metabolism: temperature 96.8–102.2°F (36–39°C).

Respiration rate approximately 15/minute.

Pulse approximately 80/minute.

Dental formula: I3/3, C1/1, P4/4, M3/3.

Breeding: usually seasonal in the wild; in the late fall in European Wild boar (Sus scrofa). Domestic and intensively reared sows capable of 2.2 and more reproductive cycles per 12 months. Sow comes into heat in 21-day cycles for 48–72 hours. Gestation 114–116 days. Litter size 6–16.

▶ **Pork and lard breeds.** (1) Pietrain boar scratching (Belgium). (2) Andalusian (Spain). (3) Poland China boar (USA). (4) Duroc boar (USA) courting. (5) Berkshire gilt (England). (6) British Saddleback sow (England). (7) Craon sow (France).

THE excitement and courage of the Wild boar hunt are legendary. The fourth labor of Hercules was to capture a ferocious boar, and as a young warrior Odysseus was scarred while killing a pig. Boar hunting was a sport of the nobility in medieval Europe. A law of William the Conqueror decreed that any commoner guilty of killing a Wild boar would have his eyes put out—an early attempt at conservation—but this animal became extinct in Britain in the 16th century, and it became much less common in all countries as forests gave way to pasture and field.

In contrast to this exalted and noble wild quarry, domesticated pigs in medieval Europe were objects of contempt because they fed on wastes—woodlots and pastures were too valuable to be subjected to their destructive grubbing. Yet they were at least partly derived from the Wild boar.

Pigs do not lend themselves to herding by nomads. Their domestication came only with permanent settlements (about 9,000 years ago). In the Orient domestic pigs took different forms in the Malaysian Archipelago, southern China and India. In Europe the Wild boar provided a large pig suitable for herding, but a smaller European variety, the Turbary pig, was more popular on account of being better adapted to life in a sty. The origins of the Turbary are unclear. It may have been derived from the oriental domestic pig, but it is doubtful that it reached Europe overland by nomadic herding—its remains have not been found outside settlements. Possibly, it migrated by means of prehistoric trading networks but a more likely ancestor is a wild pig of southeastern Europe.

Interbreeding has greatly clouded the question of pig origins. The emergence of intermediary forms, both of domesticated and wild pigs, has produced a complicated collection of animals with diverse characteristics.

There are many references to pigs and pig keeping in medieval manuscripts, but not to specific breeds. Writers seem to refer only to a Turbary type—a small, dark, long-legged and razor-backed animal which is in complete contrast to the much larger and lighter-colored pig familiar today. It is a matter for speculation when and where this change in popularity took place, but by the 18th century a large derivative of the Wild boar was predominant.

Extremely large and taking a long time to fatten, it provided large flitches of bacon and hard fat, a useful source of energy for laborers. Smaller, earlier maturing breeds provided a leaner meat. Demand for leaner meat stimulated the introduction to western Europe, beginning at the end of the 18th century, of foreign breeds. Some were from the Mediterranean, but the principal imports were oriental.

There are many systems of pig keeping throughout the world, because the pig is a very adaptable animal and can thrive on a wide range of different feeds. Highly intensive systems of production are common in Europe and North America. In these, all aspects of the pig's life are under strict control, from nutrition, breeding and health, through to environmental conditions. The objective of intensive production techniques is to optimize biological performance, often in units of considerable size. Understanding of the pig's nutritional requirements, together with evaluation of the quality of feeds, has allowed the formulation of concentrate diets which promote liveweight gains of up to 28oz (800g) daily. In Europe diets are based upon grain, usually wheat and barley, as the source of energy and

6

combinations of animal proteins—eg fish meal—and plant proteins—eg soy bean meal. North American pig diets include far more corn.

There is increasing interest in the use of less conventional and cheaper feeds. The pig has long been regarded as a useful converter of wastes into meat. Pigs were once considered an essential component of the well-managed dairy, being fed on whey and other by-products, and as early as the end of the 18th century in England, several thousand pigs were fattened annually on residues of beer production. Waste products such as these are being incorporated into modern intensively-reared pig diets, particularly in farms equipped with liquid feeding systems.

Reproduction is also closely controlled on intensive pig units. Gilts (maiden pigs) are often mated as early as 200 days of age. Natural service is still common, but artificial insemination is popular. The gestation period of around 115 days is followed by a lactation of about 21–25 days. The young are then removed and the sow may be remated within eight days. Thus approximately 2.2 or more reproductive cycles are possible within any 12 month period. At around ten piglets weaned per litter, this gives an annual output of 22 pigs per sow per year. Sows are not normally kept for more than six cycles.

Optimum environmental conditions are necessary to achieve high biological outputs. This usually means housing, to protect animals from excesses of temperature;

temperature is more critical for the new-born pig than for older animals. Housing conditions will also occasionally incorporate systems of waste disposal and automatic feeding and drinking facilities to reduce labor.

Far less control of pig production is apparent under more extensive conditions. Pigs are allowed to roam within the confines of paddocks, or sometimes completely free. A proportion of their nutrient requirements is therefore derived from plants, berries, fruits and roots together with animals including earthworms, for which the pig has a particular liking.

The pig when totally confined is entirely dependent upon man for provision of its nutrient requirements. Knowledge relating to trace elements and vitamins is not complete, and deficiency symptoms are occasionally encountered. Roaming pigs have access to a wide range of feeds and may thus obtain significant amounts of these elements.

Overall biological output is usually lower than with intensively reared animals due to the lack of control, but extensive production is still an attractive proposition, even where land prices and density of population would seemingly argue against it. Intensification has very large capital requirements, which may not favor its implementation in the West and which may completely preclude it in some underdeveloped regions.

The most traditional forms of husbandry continue to make a contribution. Pigs are ubiquitous in African villages, where they roam almost completely free consuming scraps, rooting and foraging.

Pig Breeds

Beginning at the end of the 18th century, the crossbreeding of native and imported oriental stock had a profound effect on the development of western pig types, but there was no orderliness in breeding policy. The result, far from the emergence of identifiable breeds, was a random mixture of individuals of varying colors, shapes and sizes. There are many references to named breeds in the early 19th century, but these are no guide whatsoever to any conformity.

Pig strains were often named after the area in which they occurred or after their breeder. Changes of location or ownership were invariably associated with changes of name. Although the Berkshire was gaining prominence in England at this time, it was not an identifiable breed. The name was used to describe above-average pigs farmed in the Thames Valley. The name was also

used by breeders elsewhere, who wished to upgrade their inferior stock—and did so by the use of a well-known name. Not until the middle of the 19th century did some conformity appear.

Toward the end of the 19th century, other European countries, in particular Denmark, became interested in pig breeding. Unfortunately for the British pig industry, where breeding policy was obsessed with points including color and "beauty," Danish criteria for superiority were carcass quality and growth performance. The foundation stock for their breed was a combination of Large White obtained from the few enlightened British pig breeders, together with the Landrace, the native pigs of Scandinavia—probably of similar origin to the Large White.

▶ **Ritual exchange.** Pigs, pearl shells and cash change hands in the gift-giving ceremonies of Melpa tribesmen in the central highlands of Papua New Guinea.

▼ **Domestic pig of Vietnam.** Oriental varieties such as this were crossbred with European pigs in the 1800s to make western strains leaner.

In Britain there has now been a full circle of events. About 150 years ago, pigs were of no particular breed. Individuals were selected and fattened on the basis of local demand, be it for small pork joints or large bacon flitches.

From this diverse pool particular breeds developed, characterized by specific points including color, shape and size, the criteria for selection often having no economic value. These pure breeds are regulated by breed societies, which only register those males conforming to the breed criteria, which may be modified following the opinions of the show judges. Petty inter-breed rivalry has often appeared. Currently the emphasis has moved back towards carcass and performance through hybridization, although this is not to say that the pure

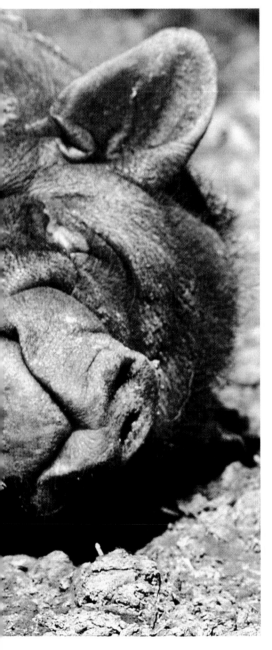

breeds have nothing to offer. They often possess specific characteristics which are or could be of considerable use in breeding programs.

Breeding trends today are similar worldwide in intensive pig production. The selection of improved types uses criteria such as growth rate and carcass weight. An early example of the new trend towards hybridization is the Lacombe, which was bred in Canada from crossings involving the Danish Landrace, Chester White and Berkshire. More recently breed companies have developed synthetic lines including the Camborough, Cotswold, and the Beltsville 1 and 2.

Southeast Asian intensive producers still rely largely on the Large White, Landrace and Duroc. Extensive pig production (other than outdoor production in the west) is not governed by any policy of nutrition or selection.

Pigs are classed into pork, lard and bacon breeds. Pork tends to come from an animal slaughtered at between 200–220lb (50–70kg) liveweight. A pig at the upper end of this liveweight range is often referred to

as a "cutter." The weight is achieved in around 3½ months under intensive conditions. Pork breeds are therefore early maturing: the laying down of fat proceeds at a young age and low bodyweight.

Bacon, which is cured pigmeat, requires a heavier carcass, as the curing process traditionally uses fairly large sides or flitches. Bacon is produced from animals of around 200–220lb (90–100kg) liveweight. An early maturing breed would be excessively fat at this weight, and therefore a late maturing breed is used.

For both bacon and pork production, strict carcass quality controls are imposed in some countries, which regulate fat content. Heavy hog production is occasionally practiced. Animals are slaughtered at around 265lb (120kg) liveweight.

It is often difficult to accurately categorize breeds. A middle maturing breed may be suitable for both pork and bacon production. Meat animals are frequently crosses, bred from two separate lines. The classification given overleaf is not definitive. Production levels of individual breeds depend on management and feeding.   JWi

PIG BREEDS

Sizes of pigs are not given since this depends on age at slaughter and nutrition rather than breed; mature body weight is almost never achieved under present farming conditions. The breeds listed here mainly reflect controlled breeding programs to produce uniform types of pig for intensive farming; uncontrolled mating under extensive/village conditions produces extremely varied local populations with individuals varying in shapes and sizes.

Pork and Lard Breeds

Berkshire

Worldwide. Originated England, with Neapolitan or Black Chinese influence.
Color pattern: black; white tips to tail, feet and ears.

British Saddleback

England. Amalgamation of Essex and Wessex, which were recognized separately from 1918–1967.
Color pattern: black with white or pink saddle around girth. Hardy with good mothering ability.

Hampshire

USA. Derived from early form of British Saddleback imported in 1825.
Color pattern: black with white saddle and white forelegs. Rather long head with straight face.

Duroc

USA. Unclear origins. Improved recently with Tamworth.
Color pattern: uniform dark russet/brown.
A productive lard breed.

Da Min

Mainly farmed and originated: China.
Color pattern: black with wrinkled skin.
Current interest due to large litter size.

Taoyuan

South central China; widely farmed in China.
Color pattern: black or very dark gray; skin wrinkled with folds; lop ears. Has high fat content, characteristic of some Asiatic breeds.

Andalusian

Farmed in Mediterranean, especially Spain; originated in Spain from Iberic type pig.
Color pattern: whitish with occasional dark patches.
Little economic value in intensive systems, but still found under extensive conditions.

Craon

Farmed and originated: France, especially Mayenne. Probably related to similar N European Celtic pigs.
Color pattern: white. Large pig with drooping ears.
A popular breed in its locale.

Poland China

USA. Originated in Ohio from diverse origins including Berkshire.
Color pattern: similar to Berkshire.
Large breed. Good lard producer.

Pietrain

Farmed and originated: Belgium. Probably related to similar N European Celtic pigs.
Color pattern: off-white with dark or red patches suggesting Tamworth connection. Characterized by large hams.
Used in crossbreeding to impart this quality.

Bacon Breeds

Large White

Large White or Yorkshire
Worldwide. Originated N England mid-1800s from Old English with Cantonese influence.
Color pattern: white.
Premier British breed exported extensively as foundation stock for American Yorkshire and European Large White breeds.

Danish Landrace

Denmark. Developed late 1900s with Large White influence.
Color pattern: white.
(The term Landrace does not refer to a specific breed but is a generic name applied to native pigs of northwestern Europe.)

Gloucester Old Spot

Originated: Gloucestershire, England. Probably related to Old English.
Color pattern: white with one or two black spots on back. Appearance has changed markedly even during this century.
Regional breed once regarded as particularly useful in consuming fallen apples.

Tamworth

Originated: Staffordshire, England. Derived from Old Berkshire with possibly Caribbean or Indian influence.
Color pattern: uniformly ruddy.
This breed has remained unchanged for perhaps longer than any other due to its unpopularity. Almost extinct, but recent revival due to its hardiness and ability to survive inclement conditions.

Large Black

Farmed: S England. Originated: Cornwall and Devon with East Anglian influence.
Color pattern: black.
A minority breed reputed to provide meat of superior taste.

Welsh

Farmed and originated: Wales.
Color pattern: white.
Of common origin to other white breeds of northern Europe, some consider this not to be separate from them.

Lacombe

Farmed and originated: Canada.
Color pattern: white.
Hybrid pig developed from crossbreeding the Danish Landrace, Berkshire and Chester White. Used widely as commercial animal in country of origin.

World Importance of Pigs

Pigs provide more meat than any other animal. Leather is an important secondary product as are bristles. A quite different function is for truffle hunting in the Perigord region of France. Truffles grow underground near oak trees and, it has been suggested, may contain a naturally occurring sexual pheromone to which sows are extremely sensitive. Pigs are also useful in clearing undergrowth, although this is probably only of historical importance. Their use as a draft animal is extremely unlikely, although it has been recorded.

Although pig meat is found worldwide, there are peoples who are forbidden on religious grounds to consume it. United Nations Food and Agricultural Organization figures indicate that there were around 780 million pigs in 1981, and just over 55 million tons of pig meat were produced. Almost half these animals were in Asia, a large proportion of them very small pigs reared on table scraps by the same families which consumed them.

In some regions of Asia feral and wild pigs are often the most abundant source of meat. Although most of the pigs are derived from escaped domestic stock, some separate wild species also contribute, such as the Bearded pig (*Sus barbatus*) from the Philippines, Borneo, Sumatra and Malaya, and the babirusa (*Babyrousa babyrussa*) from Sulawesi, which has potential as a domestic species since it is easily tamed. The Celebes wild pig (*Sus celebensis*) appears to have been domesticated in very restricted areas of Southeast Asia, where it is maintained as a village or household animal. Hybridization of this species, and others, with the domestic pig has resulted in a wide range of pig forms in Southeast Asia.

The term pigmeat describes a wide range of products, each requiring a specific type of animal and process. Whole suckling pigs are sometimes consumed on special occasions but most meat is in the form of pork from lean animals maturing at about $3\frac{1}{2}$ months, or bacon. The carcasses of heavy hogs are either cured for markets requiring large bacon joints, or is processed into sausages, salami, charcuterie and other cooked meats. Animals from breeding stock eventually find their way into pork pies and frankfurters.

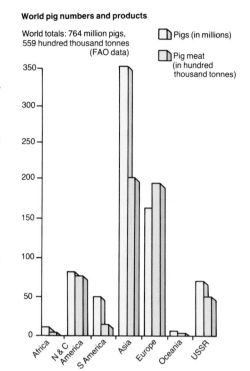

World pig numbers and products

World totals: 764 million pigs, 559 hundred thousand tonnes (FAO data)

☐ Pigs (in millions)
☐ Pig meat (in hundred thousand tonnes)

◀ **Bacon breeds.** (**1**) Landrace sow (Denmark). (**2**) Pregnant Tamworth sow (England) building nest. (**3**) Welsh sow (Wales). (**4**) Gloucester Old Spot gilt (England) clearing fallen apples. (**5**) Large Black boar (England) recoiling in defense. (**6**) Large White boar (England) in offensive threat display.

DEER AND REINDEER

FARMING deer under modern livestock management systems is a recent innovation, but the practice of using deer as a source of food and by-products is extremely ancient.

One of the major advantages of the deer family as a source of animal protein, is that different species have adapted to a wide range of conditions, very often in countries where the environment is not suited to cattle. Prior to the development of agriculture, tribes in the subarctic regions relied on the herds of reindeer (see below) for almost all their needs, and followed these herds on their annual migrations. Even after agriculture gave rise to city states, deer were one of the major sources of animal protein, and archeological evidence provides increasing indications that farmed deer were a primary source of meat long before the present traditional livestock became common.

In many European countries deer were the main source of meat through the winter and though largely left to survive in their natural habitat, men had developed enough control over herds to be able to bring them into special areas for slaughter. The sport of deer hunting had its basis in the practical need to fill the larder.

The provision of meat during winter was essential to the maintenance of a fighting force, and the ownership of deer became a prime prerequisite to power. Remnants of herding facilities used at least 1,000 years ago are still to be found in such places as the Fasque estate of the Gladstone family in Scotland.

Deer were only displaced when it became possible, with the development of winter crops, to carry fat cattle into the winter. Deer lost their commercial status but remained as a symbol of class; by law in England no commoner could hang a 12-point antler on the wall of his cottage. In Europe generally, hunting became elitist and a host of laws were introduced, some of which still are in force, severely curtailing deer hunting and creating privilege.

Farming Deer in New Zealand
Red and Fallow deer were introduced in the late 1800s to Australia and New Zealand to provide traditional sport for the upper classes of their new society. In New Zealand they found such favorable conditions, and a total lack of any natural predators, that they multiplied at an alarming rate. Unlicensed shooting of stags was eventually allowed, but it was too late. By the 1930s the deer population had reached such numbers that, though no population count was possible in

the wilderness which the herds inhabited, it was competing with the sheep and causing soil erosion through destruction of the native flora. A government campaign of slaughter was introduced, and between 1931 and mid-1967 over a million deer had been killed by the professional hunters. In the peak year, 1939, these professionals shot 41,000 deer, but they could only slow the rate of increase.

After World War II, entrepreneurs realized the potential for export of venison from these feral herds to the booming West German market. The trade became so profitable that jet-boats and helicopters were brought in to maximize slaughter efficiency. It was clear that numbers would decline so rapidly that the trade would diminish and processing facilities would be under-utilized. The pioneers of deer farming urged government to approve fencing of deer to provide a base breeding herd that could be culled regularly. The regulations were introduced in 1970 and a few farms established, but few farmers took the idea seriously.

The outlook changed when Korean importers of velvet antler, a major ingredient in oriental medicines, saw the potential value of New Zealand's farmed herds as a reliable source of quality antler. The high prices they paid created such a keen interest in deer farming almost overnight that helicopter operators stopped shooting deer and turned to live capture. Tranquilizer dart guns were used, plus "bleeper" devices so that deer could be tracked electronically, and later a gun that threw a net over the fleeing target was perfected. In the 1979–80 season over 20,000 live deer were brought in from the hills by this means.

The prices for velvet reached $85US/lb ($185US/kg) in 1979 and this, plus government incentives, produced massive injections of city capital into deer farming syndicates. The price of hinds rose to a peak of $2,500US each. The industry has now achieved a more stable rate of growth and farm-bred stock now comprise the bulk of the national herd.

The deer, especially Red deer, have adapted easily and quickly to farm life. They are handled regularly in enclosed, darkened yards for drenching (to rid them of parasites), weighing, etc, and can be grazed efficiently behind electric fences. The first farms were established on hill country with low stocking potential, which meant a high unit cost for the requisite 6.6ft (2m) fences. It soon became apparent that deer thrived on improved pasture and higher stocking rate—up to 6 hinds per acre (4 per

ha)—reduced fencing costs and gave rise to marked improvement in herd performance. Hinds in mobs of up to 50 are mated to a single sire—selection and natural dominance usually coincide—and fawns are normally weaned prior to mating of hinds.

At first deer were thought capable of surviving the winter on minimal feed but this was soon shown to be false. If anything they require better winter feed than cattle or sheep since, in their early years, they do not lay down fat reserves to carry them through a winter. Hay, silage or crop is therefore fed. Nevertheless deer are very efficient grazers and processors of ingested foods, and have a better conversion of feed into meat than sheep or cattle.

Although by 1983 velvet prices were down to $25US/lb ($55US/kg) many farmers continued to retain herds of mature stags for this trade. But there was increasing interest in carrying stags through to 15 or 27 months for slaughter; an export trade in farmed venison had been established. Regulations require that farmed deer be slaughtered in licensed premises and a number of DSPs (deer slaughter premises) had been established. Farmers were receiving about $1.00US/lb ($2.25US/kg) net, clean carcass weight, and these returns were bringing substantial interest among sheep farmers faced with declining profitability.

West Germany is the major market for game and the previous outlet for New Zealand's feral venison. But with licensed slaughterhouses where both ante- and post-mortem inspection is possible, New Zealand exporters can also sell in markets such as Japan where game is not normally admitted. With the limited quantity of farmed venison available, however, exporters were mainly confining their activity to restaurants and hotels in the more lucrative markets of North America and Australia.

Farming Deer in Other Countries

In Australia feral deer have existed in pockets in most states—Red deer in the Brisbane Valley, Fallow deer in Tasmania and other regions—but numbers are not great and no one species has dominated. Some farmers prefer Fallow deer, some Red deer, and some Rusa deer; the majority of newcomers have a few of each. Unlike New Zealand's the Australian industry is not much concerned with venison exports, having substantial markets on its own doorstep. Velvet production is less important and less handling of the deer is undertaken.

▲ **Deer farm in Western Scotland.** Medieval deer parks provided more meat than did grazing that was given over to cattle. After centuries of decline, deer farming was revived in the 1970s to make a productive use of poor land. Venison has proved so profitable that now most herds are kept on good-quality pasture.

► **Feeding eland in Tanzania.** Zebu cattle, the main agricultural meat producers in Africa south of the Sahara, were introduced in large numbers only 1,300 years ago. Native species are more fully adapted to African conditions.

Farmed species of deer

Red deer
Cervus elaphus
Red deer, maral, hangul, shou, Bactrian deer or Yarland deer

Farmed: New Zealand, Australia, Europe, Korea, Wild in Europe, N Africa, Asia Minor, Tibet, Kashmir, Turkestan, Afghanistan.

Fallow deer
Dama dama

Farmed: New Zealand, Australia, Europe. Wild in Europe, Asia Minor. Iran.

Sika deer
Cervus nippon
Sika or Japanese deer

Farmed: Asian countries and Australasia. Wild in Japan, Vietnam, Formosa, Manchuria, Korea, N and SE China.

Rusa deer
Cervus timorensis
Rusa or Timor deer

Farmed: Australia, Papua New Guinea. Wild in Indonesian Archipelago.

Wapiti
Cervus canadensis
Wapiti or elk (elk in N America only)

Farmed: New Zealand, Korea, N America. Wild in western N America, Tien Shan Mts to Manchuria and Mongolia, Kansu, China.

Mule deer
Odocoileus hemionus
Mule or Black-tailed deer

Farmed: N America. Wild in western N America, C America.

White-tailed deer
Odocoileus virginianus

Farmed: N America. Wild in N and C America, northern parts of S America extending to Peru and Brazil (introduced to New Zealand and Scandinavia).

Reindeer
Rangifer tarandus

Farmed and feral populations: China, Russia, Scandinavia, N America, Iceland, Greenland, Scotland, South Georgia.

Interest in commercial deer farming along New Zealand lines is evident in a number of countries such as India, Malaysia, Indonesia, Japan, Taiwan, and mainland China. In some of these countries experimental projects have been established and are being tested.

By the mid 1950s, British deer parks had declined to fewer than 300 from their zenith in the medieval period, when it has been suggested that they numbered well over 2,000.

A similar fate befell the parks of other European countries. But European deer farming began anew in Scotland, in the late 1960s, when government money was provided for research into the potential for improved land use in the Highlands through the exploitation of Red deer. Private deer farms were established in Scotland and England with Ministry of Agriculture grants becoming available in 1978. By 1984 an estimated 200 farms were in operation. Most of the newer units are becoming established on the best available grassland. Deer are now being slaughtered in abattoirs and marketed through a producer's cooperative.

The harvesting of antler velvet has never been practiced on any scale in Europe and was in fact outlawed in Britain on humanitarian grounds in 1978. Venison production, however, has proven financially worthwhile. In West Germany in the 1970s research was concentrated on the use of Fallow deer for meat production and a very large number of small private Fallow deer units have since become established.

Deer farms now also exist in France, Italy, Switzerland and Sweden with especially rapid growth in Holland and Denmark. The market for venison throughout Europe appears very strong with consumers prepared to pay a premium for meat from farmed deer killed at around two years. Wild deer are often slaughtered when much older and much less tender.

In North America the farming of deer for venison is almost unknown. However, demand for deer meat is increasing very rapidly and there is little doubt that deer farming will develop. In the meantime several enterprises exist for velvet production from elk, or wapiti, and there are a number of large game ranching ventures to give sportsmen the opportunity to shoot deer and antelope.

In Spain, Scandinavian countries and several Iron Curtain countries, deer herds established for trophy shooting are being developed for possible venison production.

DY/JFl

Harvesting Wild Animals—Game Ranching

Interest in commercial meat production from wild herbivores gained ground in South Africa in the late 1960s and by the early 1980s well over 2,500 tons were being exported annually. The animals mainly exploited for meat production are springbok, impalas, kudus and bontebok, but nyala, wildebeest, zebras and eland are also farmed. The best cuts find a ready market in Europe. Off-cuts are used to make "biltong" a dried meat for the local market. Sportsmen pay to shoot trophy heads and this brings the farmer an extra return for little outlay. In 1980 an eland head was worth about $320US and a sable more than $600US. The farmer himself often acts as guide for an additional fee.

Meat production from game in southern Africa, sometimes referred to as ranching, could better be described as extensive farming. Captured animals are run on country similar to their natural habitat, and considerable managerial skill has to be applied to ensure that overgrazing does not cause permanent damage to the delicate balance of plant species or cause ill-health in the animals.

It is possible to achieve a combination of different species with different grazing habits and so the gross stocking rate is much higher than could be possible with a single species. Game farming experts place great emphasis on the need to achieve the proper combination of complementary species so that the veld habitat can be utilized efficiently on a sustained basis. In some instances sheep and cattle have been successfully included.

Cropping is necessary to maintain the genetic quality of the herds. The rate of culling varies from a third in a normal year to 70 percent in a good season, but considerable skill is required in achieving the right balance of age and sex. A 1:3 ratio of males to females is considered desirable and it is important to cull at specific times of the year (these differ for each species) to minimize stress in the herds at critical times of the season.

Most of the cropping is done from a helicopter or by "spot-lighting." This is carried out by a team which includes a mobile slaughterhouse and cool truck, brought onto the farm by the game exporting firm. Farmers are paid on a dressed carcass weight and all meat for export has to be passed by government inspectors. Game farming is only slowly gaining popularity in South Africa, but live animals of the more popular species are now regularly available for sale.

The game farming example established in South Africa, like the New Zealand deer farming example, is being followed with considerable interest by various international agencies. They see great potential for increased production of animal protein for local populations in many of the less developed countries through exploitation of available wild herds and through selecting from a wide range of animal species.   DY

World Importance of Deer

In oriental countries Sika deer have been housed for several hundred years as a source of "velvet" antler—taken at the stage in growth before hardening (calcification) begins. It is used in eastern medicines. In Korea, a major importer of velvet antler, Red deer and Wapiti have recently been added to the species of deer farmed for this purpose.

A market for venison, especially in West Germany, stimulated a modest development of Red and Fallow deer farming in New Zealand in the early 1970s. There are about 2,000 deer farms in New Zealand, many running herds of 1,000 or more, with a total national herd estimated at 300,000. Wapiti are used for crossbreeding.

Other countries have since seen the beginnings of deer farming, but on a much smaller scale. Red, Fallow and Rusa deer are farmed successfully in Australia. Rusa, being native to the Indonesian Archipelago, are well adapted to tropical regions, and in Papua New Guinea a government project has been set up to exploit large feral Rusa herds. In Mauritius farming of Rusa deer is well established and closely follows New Zealand methods, while North American game parks, where wapiti (erroneously termed "elk" by some), Mule and White-tailed deer are run on extensive holdings for sporting purposes, are beginning to adopt commercial venison operations as well. Some continental European countries have small commercial Fallow deer herds, but sporting lobbies restrict development so that the Red deer are preserved for trophy shooting. British Red deer farms are few because antler cropping is not allowed and returns on more traditional forms of husbandry are high. Large numbers of top quality English park Red stags have been imported recently by New Zealand.

The total world population of farmed and feral (gone wild) reindeer is about 3 million: about 2 million in the USSR, 700,000 in Finland and Scandinavia, 30,000 in North America, 3,000 on Iceland, 2,500 on Greenland, 100 in Scotland, 2,000 on South Georgia.

In Scandinavia, Finland and the USSR about 50,000 people are dependent on reindeer husbandry for their living. In addition a number of people earn their living in slaughterhouses and as administrators in the herders' own economic associations. Worldwide, about 650,000 reindeer are slaughtered annually giving roughly 22,000–25,000 tons of meat. Most of this is offered for sale on the domestic market as steak or processed as sausage or canned products.

Although still new, the harvest of antler is already important due to its value on the oriental market. Prices vary with supply and demand but are generally considerably higher than for meat.   DY/YE

Reindeer

Laplanders are best known to the world as the people who herd reindeer, but they are not the only ones who practice this life, and they were not the first. Historians believe that attempts were made to domesticate reindeer about 7,000 years ago in southern Siberia east of Lake Baikal.

It is likely that domestication took place independently in two or more tribes. Reindeer herding then slowly spread westwards, and it is thought that some sort of management of these animals was practiced by people in northern Scandinavia around AD800. It was not until the 15th century that the modern Lapp reindeer culture was developed.

When first domesticated, the animals were probably used mainly as beasts of burden during the seasonal migrations of nomadic peoples. It is also believed that wild reindeer were tamed and used to lure other wild reindeer within range of man's primitive weapons. Females were a source of milk, a use recorded in northern Asia as early as AD499.

Reindeer husbandry is practiced today mainly in northern parts of Scandinavia, Finland and the Soviet Union although successful introductions have been made to Alaska, Canada and Scotland. In Greenland,

▲ **Antlers in velvet.** In New Zealand, some stags, such as these striking Red deer, are kept as producers of antler velvet for traditional oriental medicines. Each year a covering of velvety, fur-covered skin carries blood to the new, growing antlers. In the fall the antlers harden and the velvet dries up and sheds, unless the antlers are harvested first. Most stags are instead slaughtered young for venison.

◄ **Red deer stag and hind** in a New Zealand setting. Red deer coat colors vary through many shades of gray, brown, red and yellow. The natural range of this species is in Europe, North Africa and East and Central Asia. Herds introduced to New Zealand for recreational hunting in the 1800s multiplied uncontrollably, threatening local plant life and competing with profitable sheep until a lucrative international market for venison and velvet developed after World War II.

Iceland and South Georgia, introduced domestic reindeer have become feral (gone back to a wild existence).

Reindeer are an indispensable part of the Lapp culture. Until recently methods have developed and changed very slowly, and traditions have been passed from one generation to the next. In recent decades the numbers of people occupied in reindeer husbandry have decreased drastically and expectations for better living standards have increased. New methods have developed making the work more effective and more profitable.

The animals still hold a central position in the daily life of the people, in their traditions, language, literature and folklore. Meat and other products from the reindeer are important in daily family life, and hides, antlers and bones are important raw materials in the extensive and economically important handicraft industry.

Reindeer range over vast areas which to a great extent are unproductive for other types of exploitation. In some parts of Siberia they are still used for milk and cheese production and for transport, as pack animals and mounts, although meat and to some extent hides and antlers are the main prime products. Until some decades ago reindeer husbandry in northern Europe was characterized by intense herding and by roundups in summer and the fall for earmarking the calves, for separating mixed herds and for slaughtering. In some areas daily corraling during a few summer months was practiced for milking and to give the animals some relief at special smoke fires from pestering insects.

In modern reindeer breeding the sex ratio is usually biased considerably toward females. This guarantees a good crop of calves which can be heavily harvested before food shortages in winter become acute.

The reindeer are often fed supplements during winter, and they are kept rather than herded. Helicopters and snowmobiles are used in the roundups, specially designed trucks are used for transport to well-equipped slaughterhouses, and energy-expensive migrations are to some extent replaced by truck transport between summer and winter pastures.

Reindeer are particularly well adapted to arctic conditions by having a thick coat which insulates against cold, wet and windy weather, and a muzzle which is covered with hair. The feet are broad to aid travel in soft snow, with sharp edges enabling the animals to dig for food in deep and crusted snow.

During the summer the reindeer will accumulate considerable amounts of fat for use in winter when forage is often severely restricted. Food shortages in winter are also compensated for by considerably reduced activity, minimizing the animal's energy expenditure.

In winter, and when the calves are small during the summer months, reindeer are particularly vulnerable to predators, mainly wolves, lynx, wolverines and Brown bears. Other important killers are parasites and disease, eg brucellosis. In some districts road and train traffic kills a considerable number of reindeer, and other accidents of occasional importance are avalanches, drowning and falling from cliffs.

The expansion of human activity creates an increasing pressure upon traditional rangelands and migration routes. Disturbances during sensitive periods are increasing and grazing areas are giving way to hydroelectric development, mining, modern forestry, roads and expanding tourism. In areas where wild and domestic reindeer overlap, mixing, often with considerable losses of domestic stock, may create serious problems.   YE

BEASTS OF BURDEN

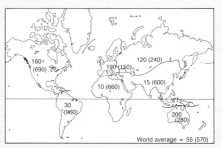
Main Species used as Beasts of Burden:

Horse
Equus caballus
Order Perissodactyla, family Equidae.

Ass or **Donkey**
Equus asinus
Order Perissodactyla, family Equidae.

Mule
E. caballus × *E. asinus*
Order Perissodactyla, family Equidae.

Oxen
Bos primigenius
Order Artiodactyla, family Bovidae.

Yak
Bos mutus (B. grunniens)
Order Artiodactyla, family Bovidae.

Water buffalo
Bubalus arnee (B. bubalis)
Order Artiodactyla, family Bovidae.

Dromedary
Camelus dromedarius
Order Artiodactyla, family Camelidae.

Bactrian camel
Camelus bactrianus
Order Artiodactyla, family Camelidae.

Llama
Lama glama
Order Artiodactyla, family Camelidae.

Reindeer
Rangifer tarandus
Order Artiodactyla, family Cervidae.

Asian elephant
Elephas indicus
Order Proboscidea, family Elephantidae.

Dog
Canis familiaris
Order Carnivora, family Canidae.

▶ **Foraging party.** With grass for a night's feed, elephants cross the Rapti River in Nepal. Although more famous for their military and ceremonial uses, these animals are used mainly in forestry work to haul heavy timber.

ANIMALS have been used for work for thousands of years and the species employed range from the dog to the elephant. There is archaeological evidence that reindeer-drawn sledges were used in northern Europe before 5000BC and that land-sledges drawn by oxen were used in Mesopotamia before 3500BC. Today beasts of burden are still an essential part of agrarian life throughout the world.

Draft and pack animals are extensively used in harsh environments where adaptation to altitude, extreme temperatures and difficult terrains are required. Some animals have become specialized to withstand these conditions by the rigors of natural selection over thousands of years. Man can hardly have improved the camel's adaptation to the desert or the reindeer's ability to survive arctic winters. Under less extreme conditions man has more scope, through careful selection, for the creation of specialist work breeds such as heavy horses and pit ponies.

Detailed information about the domestication, biology and other uses of many of these animals can be found in their main entries elsewhere in this volume. There follows here a survey of how domestic species fill their roles as beasts of burden.

The **horse** was first domesticated about 5,000 years ago in southern Russia and its use rapidly spread. In early times it made a greater impact on warfare than on agriculture. It was not until the Middle Ages that the great war horses were put to work on the land and the first steps were taken towards creating heavy breeds such as the Shire. Gradually the collared working horse usurped the yoked ox on the land, because it could plow faster on most soils. Disputes over the merits of the two animals continued until the 19th century, when the strength and endurance of the ox team was still required on heavier soils. Extensive use was also then being made of horse-powered farm machinery.

The 66 million horses in the world today belong to a bewildering variety of breeds specialized for different uses, including plowing and powering machinery, pulling carts and carriages, riding and sports. Modern draft breeds include the British Shire and Clydesdale and the French Percheron and Ardennes. The Shire is the largest of these breeds, standing over 18 hands (72in—180cm), weighing 1 ton and capable of moving a 5 ton load. In western Europe and North America the number of working horses has dwindled while in eastern Europe, Asia, Central and South America and North Africa they have retained their importance.

The **ass** or **donkey** has long been regarded as the poor man's beast of burden. Capable of working under the harshest conditions, it is one of the most widely distributed of the domestic animals. The world population totals 39 million and while use of the ass in Europe is declining, in developing areas such as Africa it is on the increase. This animal's primary role has always been as a pack or baggage animal although it is also used to draw carts and plows and to power threshing and milling machines. The ass's

The Largest Living Tractor—the Elephant

The elephant is still vitally important for logging operations in southern Asia. It can pull out and transport the valuable tropical hardwoods from dense forest where vehicles and modern machines cannot venture.

It is a magnificent hauler. A single elephant can easily pull a log of 2 tons and a pair in harness will pull 5 tons.

In the wild, elephants will push down trees with their foreheads and lift branches with their tusks and trunks in order to obtain the higher shoots. It is these natural abilities, as well as their great strength, which man exploits.

Archaeological evidence of the domestication of elephants in Asia dates back to the second millennium BC when they were primarily used for military purposes. The Persian King, Darius, used elephants against Alexander the Great, and Hannibal's march across the Alps with tamed African elephants is well known.

Today it is only the more docile Asian elephant (*Elephas indicus*) which is used for forestry. The commonest method of capturing these elephants, which are not normally bred in captivity, involves driving wild herds into specially constructed *keddahs* or corrals which enclose a part of the forest containing a stream.

Once captured the animal is securely tethered in constant view of its trainers who hand feed and stroke it. After two or three weeks it is tethered between two large tame elephants and is forced to respond to the mahout's commands by them. A *hawkus* (metal-pointed stick) is used to punish the elephant if it misbehaves and it quickly becomes tame and obedient. Eventually it is assigned to a mahout with whom it may work for 30 years.

Although the mahout does not normally own the elephant (it being worth more than he could earn in a lifetime) he spends much of his time feeding, bathing and tending the animal in his charge.

▷ **Camels dot a hillside in India** OVERLEAF as their owners gather for a festival. The number of dromedaries (one-humped camels) in northwest India, Pakistan and Afghanistan is increasing and is second only to the numbers in North Africa. Camels of the Arabian Peninsula are in relative decline.

◀ **In the late-morning heat,** Indian zebu cattle trudge before a heavily laden cart. India has over 200 million cattle and they are bred primarily for plow and cart, with dairy products second in importance. Work is also the main product of cattle in most other parts of Asia.

◀ **Donkey transport in Morocco.** BELOW Domesticated earlier than the horse, the ass has always served more peaceful purposes. For thousands of years horses belonged to the wealthy, who used them in war, while the common man fought on foot in wartime and rode on an ass to go to market in peacetime.

versatility and modest requirements make it essential for small-scale agriculture in countries ranging from Ireland to China.

The **mule** combines the endurance and sure-footedness of its sire, the ass, with the size and strength of its dam, the horse. These qualities, in addition to its hybrid vigor, enable it to carry heavier loads than either of its parents. The reverse cross is known as a hinny. There are Biblical references to mules. They were probably first bred in Asia Minor around 800BC . Mules were of great importance in the Roman Empire, for riding, plowing and drawing carts. The Romans even devised a "machine" to enable the ass to copulate with the mare more easily. They also used mules in army baggage trains, a practice which armies continued up to World War II. Today's world population of 14 million mules is on the increase, with extensive use of these animals in North Africa, Mexico and China.

The Romans, who drank no milk, used **cattle** exclusively for work and in northern Europe cattle alone were used for plowing until the Middle Ages. Pairs of cattle drawing land-sledges gave way to pairs drawing carts and plows. On heavier soils teams of up to eight animals, sometimes mixed with horses, were used. Similar teams are used today for heavy haulage in Asia.

There are now 1.2 billion cattle in the world and it is estimated that up to 40 percent of these, found mainly in Asia, are used for draft. They are by far the most commonly used work animals. Oxen (castrated males) are generally preferred, being stronger than females and more docile than entire males, although cows may sometimes be used.

Early methods of attaching the oxen to a plow or cart involved ropes tied round the horns. Later improvements linked oxen by wooden beams across the shoulders and fitted vertical bars on each side of the neck to hold the yoke, thus formed, in place. A new design of yoke to fit the *maresha* (the traditional African plow) has recently been introduced in Ethiopia. It consists of an inverted wooden V shape which enables plowing by a single animal rather than the pair required for the conventional yoke. The animal is controlled by a rein attached to a nose plug or ring which pierces the nasal partition. Oxen are also used to power mills and threshing machines and to raise water.

In Africa, zebu and Africander cattle are most commonly used for work, while in China, Taiwan and Vietnam, Chinese yellow are the predominant draft animals. All these breeds are derived from the Asiatic zebu cattle. Bali cattle, about 1 million in Indonesia, are draft animals domesticated entirely from stocks of the wild banteng (*Bos javanicus*). They are well suited to work on the small-holdings of Southeast Asia as they are easy to train, thrive under poor conditions and are resistant to ticks.

The **yak** is found in the highest mountain ranges of Central Asia stretching from the Himalayas to Siberia and including areas of Tibet, Nepal, Mongolia and China. It is well adapted to high altitudes where it can outrun horses, which suffer from the lack of oxygen, but it loses its vigor at altitudes below 10,000ft (3,000m).

More than half the people in the world depend on rice for their staple diet and the **Water buffalo** is essential for the cultivation, harvesting and threshing of the rice crop. Water buffalos were first domesticated around 2500BC in Indochina and spread to all parts of Asia. Today there are 130 million

Economic Importance of Beasts of Burden

The economic importance of draft animals worldwide is difficult to assess due to the diversity of roles which an animal may fulfil. Thus total populations of animals can be misleading, particularly those of cattle and buffalos where animals may be used exclusively for draft, meat or milk or any combination of the three. The reindeer is an integral part of the Lapp's life style. The animal's draft role is inseparable from its other many functions. This is not the case in subsistence agriculture, where the draft ox, buffalo, ass or mule is used primarily in crop cultivation, other functions being secondary.

Mechanization has diminished the importance of working animals in the developed world. In Europe there are 17 times as many tractors per head of population (0.017) as in Africa and Asia (0.001). In Africa the ass and mule (0.029 per head of population) appear to be the main source of power while in Asia the buffalo takes this role (0.045php). Australia and New Zealand are least dependent on animal power with only the horse present in significant numbers (0.03php). Not surprisingly North and South America have the highest numbers of horses (0.05 and 0.058php) reflecting the importance of the horse in traditional cattle ranching.

It is estimated that about 480 million of the world's 1.2 billion cattle are worked, making the ox the most popular draft animal, followed by the buffalo (122 million), horse (66 million), ass (39 million), camel (16.8 million) and mule (14.2 million). The total number of tractors in the world is 21.3 million which indicates that mechanization still has a long way to go in order to replace the draft animal.

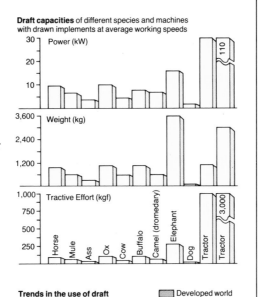

Draft capacities of different species and machines with drawn implements at average working speeds

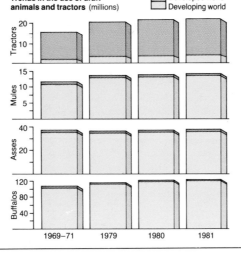

Trends in the use of draft animals and tractors (millions)
Developed world
Developing world

in the world and the population is increasing. The highest concentrations are in Bangladesh, India and Indonesia.

The title "ship of the desert" perfectly describes the historical role of the **camel**. In desert exploration and trade it parallels the role of ships in the seafaring expeditions of the Middle Ages. The camel's ability to carry heavy loads long distances through some of the harshest climates in the world was exploited in the development of the spice trade between East and West and later in the exploration of Africa and Australia.

The one-humped dromedary is found in the desert areas of Africa and southwest Asia and it can carry a load of 495–650lb (225–295kg) for up to 20mi (32km) per day. It is finer-limbed and faster moving than its close relative the two-humped Bactrian camel which is found in Central Asia, China and Mongolia and can carry 595lb (270kg) 22mi (35km) in a day. While mainly pack animals, camels are also used to power sugar-cane crushers and oil seed mills.

The world population of camels now stands at 16.7 million; 12.3 million are found in Africa, and the rest are distributed throughout Asia.

When the Spanish conquered the Incas in 1531 they found animals which had been domesticated for 4,000 years. The **llama** has retained its role of sure-footed pack animal in the mountainous regions of Bolivia, Chile, Argentina and Peru.

Reindeer are an integral part of the Lapp way of life in the Scandinavian Arctic. They provide meat, milk, bones and hide as well as draft power in winter and pack work in summer. Reindeer are better than both dogs and horses at pulling sledges over frozen ground and one animal can draw a load of 300lb (140kg) at 8mi (13km) per hour and cover 25–35mi (40–55km) in a day. In Siberia they are also used for pack and riding and a strong male can carry 143lb (65kg) for up to 50mi (80km) per day.

The Husky, Samoyed and Laika breeds of **dog** have long been used by Eskimos, Indians, trappers and explorers to haul sledges in the frozen Arctic. There are numerous Husky varieties which are distributed through the Northwest Territories of Canada and in Alaska and Greenland. They are paired in teams of 4–16 animals. Eight dogs can pull a load of over 1,500lb (6,750kg). All these breeds have long thick fur to protect them against the icy cold. A stable dominance hierarchy in the pack, with the driver at the top, is vital for the coherent, efficient and effective working of the team.   MAB

FOWL

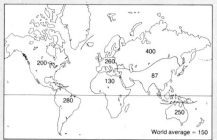
▶ **Madcapped glory of a champion,** a cock that takes the Bearded Polish prize.

ONE of the earliest uses of domestic fowl was military. The cock, crowing his pre-dawn challenge, roused the soldiers in time to position themselves for battle by first light, and he typified the courage leaders wanted to instill in their men. Portable, fear-less, almost insensitive to pain, amorous and loyal, he was the ideal mascot. The bright colors of his plumage, comb, spurs and scales, and his aggressive nature inspired designs for battle dress and armor through-out history.

Chickens also parallel other human traits. They are one of the few creatures in which male and female routinely share the family chores. The male leads the defense of the home, whose site he helps to select, and the male and female sleep together.

Historians tell us the chicken was domesticated from wild jungle birds in southeastern India and the Malay Peninsula at least 6,000 years ago. It spread west and became well established in Egypt by 2000BC. Eastward movement was taking place at the same time.

The Persians taught the Greeks cockfight-ing, and the Greeks gave it to Rome. It was the Romans who took chickens to western Europe. In many places chickens were bred for sport long before they were recognized as a primary food producing animal. They were frequently used in religious rituals.

Egypt developed artificial incubation for large numbers of eggs at the time when the pyramids were being built. This may have been a response to the problems of feeding masses of people in congested areas. The Chinese too had mastered the art of mass production of poultry before the time of Christ. It was 1,800 years before western Europe and the United States reached the same level of development.

Ducks, geese, guineafowl and, much later, turkeys were gradually added to man's domestic stock of birds. The process continues today as quail and pheasants are kept under more and more controlled condi-tions (see pp117–119).

By the 3rd century BC, the Greeks were using selection to produce specialized breeds of chickens. Plato called chickens "bipeds with feathers" and Columella, a 1st century Roman writer, made many points about their biology which have since proven to be true. He also referred to "Adriatic hens" which were very small, presumably bantams.

During the Middle Ages, chickens were offered as payment of rent and as duties to lords and feudal masters. Because the chicken was nearly self-supporting and usu-ally cared for by the tenant's wife and chil-dren, demanding several of them annually in exchange for the use of the land or living quarters was not thought to impose a great hardship.

Chickens for Meat

From earliest domestication the flesh of all of the species of birds man keeps for food pro-duction was recognized for its flavor and ease of digestion. Throughout the centuries chicken broth or chicken soup has been recommended to the sick and ailing, credited with medicinal properties. Its real worth more probably is its ease of assimila-tion by a stressed system.

Modern selective breeding of meat birds began to develop in the 1800s. At this time it was mostly art, and far from the scientific approach that was to follow, but through visual selection, improvement was made. The theory of "like begetting like" gave rise to the practice of culling, and the "eye" of the most successful poultryman was felt to "see qualities" that escaped the average per-son's view. Some farmers gained a repu-tation as producers of good stock and others gained recognition for their ability to fatten them. Thus, specialization established itself in the poultry world.

Starting in the late 1800s young birds were confined in small crates, or batteries, as they approached maturity. Feeding them on moistened grains and milk for 3–5 weeks produced a tender, moist carcass, a vast improvement over the sinewy bird that pre-viously roamed the fields in search of enough bugs and wild seeds to support itself.

Capons (young castrated males) became popular, for they were less active and had youthful flesh during a longer growing period. Fryers—or broilers, as they are now called—were a seasonal item available in late summer and early fall. Year-round pro-duction of hatching eggs in the 1920s ended their seasonability. The modern broiler industry developed into a commercial busi-ness with the constant availability of day-old chicks and the nutrition that would sup-port good growth, indoors. Broiler growing gradually became specialized but most of the production was in relatively small units of 1,000 birds or fewer on general farms until about 1940. Broilers are young chickens (about 8 weeks old) kept in confinement, growing to about 1.8kg (4lb) liveweight and producing carcasses of 2–3lb (0.9–1.4kg).

During World War II, the chicken really demonstrated its versatility as a meat pro-ducer. Able to respond quickly because of a short generation interval and early market

age, the consumption of meat from chickens advanced worldwide. It has continued to advance since then. Spurred on by such events as the Chicken-of-Tomorrow Contests in the United States in the late 1940s, broilers have reached the record weight of 4.25lb (1.93kg) liveweight at $7\frac{1}{2}$ weeks of age on feed conversions of less than 4.4lb (2kg) of feed to produce (2.2lb) 1kg of liveweight.

Capons still command a premium in many markets, but with male chickens currently achieving good market weights before sexual maturity creates social unrest or causes them to be staggy (develop hard, stringy flesh of a coarse and sinewy appearance that is typical of the male) there are fewer producers. The extra risk and recuperative time entailed by castration is often felt to be unnecessary and the severity of the operation has caused its humaneness to be questioned in some countries. From the 1940s to the 1960s various attempts were made to effect a chemical caponization by injecting males with slow absorbing pellets containing female hormones. The effect of these implants was only temporary but gained some acceptance before regulatory agencies in several countries banned their use. The feeling was that possible residual amounts of the drug remained in recently injected birds as a contaminant, and the preparations were withdrawn from the market in most countries.

The meat chicken industry is based largely on the broiler throughout much of the world today. In many countries, vertical integration has been employed to create huge complexes where breeding flocks, a hatchery, a feed mill and a processing plant are operated by a single business organization. This may involve a number of large privately owned growing houses operated under contract or they may be company owned. Interlocking arrangements make it possible to schedule a planned number of market-ready birds for processing each day and they permit economies of scale with low sales costs and good use of forward planning. While details will vary from one country to the next, quality control and product standards are prime considerations in this type of production complex. These business structures have made chicken continuously available at very reasonable prices throughout much of the world.

Meat and utility breeds of chickens. At first, although chickens were identified with certain localities, they bore names based on their characteristics, their appearance, or the way they were raised. In the 1700s, early in the period of agricultural reform in Europe, certain regions attracted the attention of buyers. Regions became more important in the naming of varieties of fowl, and farmers began to select their stock to fit a local image. The Surrey, Sussex and Dorking fowls were examples of this in England as were the Rhode Island Red in the United States and the La Bresse in France. This was the start of commercial poultry breeding.

Strains were greatly improved as poultrymen learned more and more sophisticated principles of selection. Some breeds, for example the Sussex, Rhode Island Red and Plymouth Rock, are also

▶ **Breeds of chickens.** (1) Wyandotte (USA: dual purpose). (2) Welsummer (Netherlands: dual purpose). (3) Ancona (Italy: layer). (4) La Flèche (France: dual purpose). (5) Plymouth Rock (USA: dual purpose). (6) Campine (Belgium: layer). (7) Rhode Island Red (USA: dual purpose). (8) Leghorn (Italy: layer). (9) Barnvelder (Netherlands: dual purpose).

World Importance of Domestic Fowl

An estimated 7 billion chickens contribute to the human diet. Many millions in parts of Africa and Asia are semi-wild. They lay barely more eggs than are needed to reproduce before they meet a natural death or make a tasty stew marking a special occasion in a mainly vegetarian diet.

Where improved breeds and modern husbandry are in use, the bird's role is very different. In 1982 the United States produced 4.4 billion broilers—meat chickens grown in confinement until about eight weeks old, weighing up to 4.4lb (2kg) live and dressing out as carcasses of about 2.75lb (1.25kg). Brazil produced 1.1 billion, Japan 879 million, France 666 million and the USSR 634 million. Millions of roasters—chickens from 3–6 months old with a dressed weight of 4.5lb (2kg) or more—found their way from commercially operated farms to a mass consumer market. Adult stewers, usually females no longer useful for egg production, also made a heavy contribution to the table.

In 1983, 1.85 billion specially bred hens layed 380 billion eggs in controlled environments. The USSR had the highest laying population (316 million) followed by the United States (280 million), Japan (123 million), France (74 million), Mexico (67 million) and Italy (63 million).

Such economies of scale have not been generally applied to ducks, which produce a minor proportion of the world's meat and eggs. Duck carcasses are high in fat (28.6 percent) and like goose carcasses (at 31.5 percent) have lost favor on account of a rising concern over excessive calories in the diet.

Fat, abnormally enlarged goose livers continue to be of economic importance, especially in France, where they are produced by force feeding young geese on a high carbohydrate diet. The livers are made into pâté de foie gras. Feathers are an important by-product of ducks and geese, as of all domestic fowl.

Turkeys are meat birds, traditionally reserved for festive occasions because of the large size of the carcass. In countries with large-scale production, for example the United States, Italy, France and Britain, consumption between holiday periods has been boosted by the retail sale of turkey parts, luncheon meats, turkey sausages and hot dogs, precooked turkey rolls and TV dinners.

In 1981 the highest per capita consumption of poultry meat was in Israel (34.7lb—15.8kg), followed by the United States (29.5lb—13.4kg), Romania (26.3lb—11.9kg), Australia (21.8lb—9.9kg) and Spain (20.6lb—9.4kg).

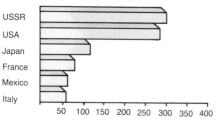

Poultry meat consumption per capita 1981

Israel / USA / Romania / Australia / Spain

kg annually

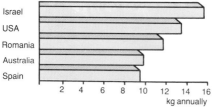

Laying hen population (estimated) in millions

USSR / USA / Japan / France / Mexico / Italy

World total: 1,850 million

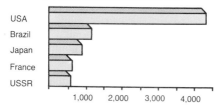

Broilers produced (estimated) in millions, 1982

USA / Brazil / Japan / France / USSR

known as "utility" breeds because they are capable of laying appreciable numbers of eggs as well as having a meaty carcass.

Few purebred birds are now used for meat. The term "breed" refers to a group of individuals possessing similar characteristics (size, shape, etc) which when mated with each other produce offspring with those same characteristics. Today most meat production is by chickens which result from crossing different breeds. They would not breed true if mated but their hybrid vigor makes them hardy and makes them grow rapidly.

Many of the hybrid meat strains are made from female White Plymouth Rocks and male lines carrying more or less Cornish breeding. From the mother they get rapid growth, rapid feathering and efficient feed utilization, making maximum weight gains on minimum amounts of feed. From the father they get their broad heavily-muscled bodies, tenacity and competitive spirit. In some parts of the world, a female line with a white skin is used to better suit local market conditions.

Most of these special meat type birds are sold under franchise arrangements where centralized breeding farms offer their stock through area hatcheries or sales representatives. Such names as Ross, Pilch, Peterson, Lohmann, Shaver and Arbor Acres are but a few of the prominent lines. In general, all of these birds have white feathers for a cleaner appearing carcass when slaughtered.

Many breeds are used privately, or on a localized basis for meat production. These breeds include: Barnvelder, Rhode Island Red, Jersey Giant, Houdan, Australorp, and Orloff. Because of slower growth rates, dark-colored pin feathers and lesser feed efficiencies, none of these compare very favorably with the specially developed meat crossbreds. Exceptions to this can be cited in such cases as the La Bresse fowl in France. There a strain with white feathers is handled in a particular way and has gained favor in the marketplace.

Eggs for Food

Eggs (see pp124–125) are symbolic to man of the beginning of life. Although frequently included in fertility rites and celebrations such as Easter, they did not become commonplace as a dietary item until after AD1800. People preferred to keep eggs to hatch the next generation.

While artificial incubation of eggs had been practiced in China and Egypt in ancient times, it was the incubator as described by de Remur in 1750, and made commercially available in the 1840s, that really set the stage for widespread mass poultry culture.

This device, with accurate temperature control, made hatching a science instead of an art and permitted man to build an industry based on the reproductive potential of the fowl. As the science of genetics emerged from the work of Mendel and Darwin, birds proved to be the most responsive animal to breeding techniques. Eggs soon far exceeded the numbers needed for reproduction and became a major food item.

Some chickens, particularly those developed in the Mediterranean area, and some ducks responded overwhelmingly to selection for egg production, but production was seasonal. The domestic birds, like their wild ancestors, selected a nest and began egg production in the springtime in response to increasing length of day. They went broody, reared their families during the long days of summer, molted during the shortening days of the fall and rested during the winter. In the 1890s it was learned that artificial light could be used to lengthen the hen's working day. This pointed towards year-round egg production. However chickens reared in the absence of natural sunlight were prone to rickets. After it was discovered at the University of Wisconsin in the 1920s that vitamin D prevents this disease, poultry production indoors became a reality.

Today several countries record average annual egg production of over 200 eggs per hen per year. Some flocks have exceeded 270 eggs per hen on a yearly average for their thousands of hens, and individual hens have passed the 300 egg mark. Individual records of more than one egg per 24 hours over many months have been achieved. These high numbers are achieved only when the female is provided with an environment that is nearly devoid of the stresses of competition, disease, extreme temperature variation and the influence of seasonal changes in day length. In addition, free and easy access to water is essential and the diet must meet the needs of both body maintenance and egg formation.

Egg-laying breeds of chickens. Before recent times, there was little use for special egg-laying breeds of poultry. Flocks were small and what little surplus they produced after ensuring a next generation was only a barter item of marginal economic importance. But towards the end of the 19th century large-scale poultry farming became possible. It occurred in California because of

a mild, dry climate. It became an option in other climates with the development of artificial incubation and confinement housing systems. Flocks with good egg-laying records gained attention and stock buyers sought them out.

One poultryman whose birds were in demand was Tom Barron, in England. They were White Leghorns, originating in the Mediterranean area. He selected for large body size and good numbers of large eggs. It was possible to breed out the hen's tendency to go broody—to sit incubating her eggs in a torpid state—thus releasing her for continuous egg production. Some meat birds are also good layers, eg White Plymouth Rock, Sussex and Wyandotte.

In the 1930s poultrymen began crossing breeds as well as strains within a breed to take advantage of hybrid vigor. It was in the late 1930s that Henry A. Wallace in the United States began applying to egg layers the system that had made crosses of inbred lines so successful in corn breeding. Thus, the "Hy-Line" strain was created (a four-way cross of inbred lines) and remains today one of the world's leading strains of egg production stock. Many other breeders followed suit and their names today rival Hy-Line in the world's markets. These names include: DeKalb, Hi-Sex, Shaver, Euribrid, Ross, Babcock, H & N and a host of others. The birds are extremely efficient converters of feed to food, often producing more than a dozen eggs on less than 4lb (2kg) of feed.

▲ **Misty morning in Bresse,** eastern France, where rearing poultry is a main industry. Here a small flock of free-range birds is fed grain. Egg numbers from purpose-bred layers such as these have been greatly increased by breeding out the hen's natural tendency to go "broody," her tendency to stop laying after 10–15 eggs and give all of her attention to incubation.

◄ **Breeds of chickens** CONTINUED. (10) Sebright Bantam (England: ornamental). (11) Nankin Bantam (England: ornamental). (12) Rosecomb Bantam (England: ornamental). (13) Sussex (England: dual purpose). (14) Dorking (England: meat). (15) Derbyshire Redcap (England: ornamental). (16) Old English Game (England: ornamental and cock fighting).

Ornamental Fowl

Quite a different function of domestic fowl is their ornamental one. Chickens and their miniature cousins, the bantams, are very good hobby animals. They are not physically demanding, pose very few threats to human health, are economical to purchase and maintain, adapt to a variety of living conditions and exist in a wide array of colors and forms. Surplus and less than desirable individuals as well as eggs not needed for hatching make a welcome addition to the family table.

The care of ornamental and exhibition poultry is also good therapy for busy people. It requires that for a few moments each day, concerns for other things must be set aside and attention focused on the birds who come to expect this and react to their owner's presence.

Perhaps the greatest drawback to the keeping of exhibition poultry is the loud and frequent crowing of the males in the early morning hours. This habit has led some communities to ban all poultry from residential areas. An outright ban is unfortunate as quarters can be designed that reduce the noise to acceptable levels.

Breeds of ornamental fowl. The great diversity in shapes, colors, sizes and temperaments in chickens has intrigued animal breeders and hobbyists and fostered the development of the poultry show. In the showroom a type of sport exists among the fanciers who breed the various color patterns and physical conformations ever closer to a designated ideal.

The competition is equally as keen as that found among athletes, jockeys, or race car drivers. It provides a challenge that is resolved in open competition. Winning represents recognition of the bird produced and the person who bred, trained and conditioned it.

Those who evaluate the birds and place the awards are extremely knowledgeable individuals who have spent years in studying and usually breeding various ornamental breeds according to the accepted breed standards. Judges are guided in their decisions by books of standards. These may vary slightly from country to country but all are definitive in their description of each breed and variety's size, shape, color and physical features.

Most of these standards had their beginnings in the mid 19th century as it became clear that the variations in birds were becoming so great as to make a systematic classification necessary. Foremost among these has been the American Standard of Perfection, the British Poultry Standards and the Dutch Pluimvee—Standard NHB—DV. These standards are modified from time to time as breeders fine tune their points of emphasis.

Included in the ornamental fowl category are miniature chickens—known as bantams. Bantams have been bred in the miniaturized likeness of nearly every breed and variety of the normal-sized chickens. In addition, there are several that have no normal-sized counterpart. The most notable of the bantam-only breeds are the Sebrights, Silkies, d'Uccles (Booted), D'Anvers and Rumpless. In general, bantams are about one-fifth the size of their larger namesakes and possess the same features, colors and shapes.

Ducks and geese are also bred for their exhibition qualities; and many poultry shows provide classes for them. The smaller breeds of ducks such as Calls, Black East Indias and Runners have found the most favor among show people and fanciers.

polygynous: the male mates with several females.

Most domestic ducks have been developed as selections from the European Mallard. The exception to this is the Muscovy which is a South American native. Some question also exists about the Pekin, which was developed in its modern form in China and spread to the western world. It is anatomically similar to the other domestic breeds and freely crosses with them, producing fertile offspring; it may well have descended from Asian versions of the universal Mallard in parallel to the other breeds in Europe.

The **goose** has been a farmyard companion of man for many centuries. Since

Ducks and Geese

Ducks and geese, domesticated around 500BC, are known collectively as waterfowl and differ from other domestic poultry in both appearance and habits. The feathers of both ducks and geese are very numerous, cover the entire body and head and form a thick and effective insulating blanket. It is this quality that attracted man to select the smaller body feathers and underfeathers (down) of ducks and geese to fill bed quilts called "feather ticks" and to incorporate them into the lining of clothing.

Although first practiced centuries ago this use of feathers continues today in many parts of the world. The demand for feathers was so great that the practice of live plucking of geese became commonplace in the 17th century and continues today in some areas. By this practice feathers are removed from the body of the goose two or three times each year with only momentary discomfort to the bird.

Both geese and ducks produce a flavorful carcass with most of the edible meat located on the breast and thighs.

Ducks will spend much of their time in the water if given the opportunity but most produced commercially in the west are raised without swimming facilities. All ducks must have adequate water for drinking and will consume large amounts. In the wild many ducks feed in the water and domestic ducks often attempt to transfer food into their water before consumption. This practice is very messy and wasteful as much food is lost in the process. Some ducks, particularly in oriental countries, are raised on liquid rations.

The voice of the drake is soft and subdued while the female makes a harsh and raspy quack. The exception is the Muscovy where the male utters a hissing noise and the female normally makes little noise at all although it is capable of a guttural "aawwk." All domesticated ducks are it has a long life and an independent nature, it has figured in literature, folklore and art and supernatural powers are sometimes attributed to it.

Geese are ever on the alert. Possessed of good vision and keen hearing, they are sometimes used as guards since it is virtually impossible to approach a goose without its being aware. Certain breeds (the Chinese and African in particular) almost automatically utter a loud cry ("honk") upon the approach of a stranger. They cannot be silenced or easily subdued as they remain aloof while sounding their cries of alarm.

Geese enjoy swimming but feed principally on land. Having voracious

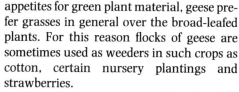

◄ **More successful than the chicken** in wet, peasant farming conditions, 75 percent of all domestic ducks are found in southern and eastern Asia. They are more disease resistant then chickens and forage better. These ducks on the island of Bali have an upright posture that allows them to walk long distances rather than waddle as they are herded from rice paddy to rice paddy as weeders.

▼ **Foster-brooding the common duck** is a typical use for a Muscovy, for she is more efficient at incubating the eggs than the ducklings' natural mother. Native to South America, the Muscovy duck is an important domestic animal in Southeast Asia, and also in western countries, especially France.

appetites for green plant material, geese prefer grasses in general over the broad-leafed plants. For this reason flocks of geese are sometimes used as weeders in such crops as cotton, certain nursery plantings and strawberries.

Although geese are considered polygynous, they are often quite selective in their choice of mates. This sometimes creates problems for people attempting to breed geese. The larger breeds, such as the Toulouse, have been developed for size and body conformation to the degree that fertility is often rather poor. Coupled with a relatively small number of eggs, which are often difficult to hatch, low fertility has caused the price of day-old goslings to be quite high.

At times it is impossible to obtain the number of goslings desired for commercial operations. Geese have not responded to man's selection for increased egg numbers the way chickens, ducks and turkeys have. Therefore geese remain better suited to small, family farm flocks than to the huge commercial scale of production commonplace in the other species.

Geese are capable of very rapid growth during the early stages of their lives. Weights of as much as 7lb (3kg) at five weeks of age have been reported. This sometimes leads to fatigue and problems of leg support. It really means that young geese should be grown in small groups of their own age as it may be difficult for them to compete in large flocks.

The practice of force feeding young geese as they approach maturity is called "noodling." Finely ground grains are moistened and placed in a sausage stuffer to make huge noodles. They are then plunged into warm water to make them slippery and placed into the goose's open mouth. The process is repeated several times each day for six to ten weeks. This high carbohydrate diet results in a greatly enlarged liver which is used to make pâté de fois gras.

Geese are very social animals and keeping one in an isolated situation can be quite traumatic for it. At an early age, these birds "imprint" (transfer their species preference) very easily and can make interesting pets. A strong bond can develop between the bird and humans. It should be remembered that the goose is a long-lived animal and such an arrangement implies a long-term commitment as the bond is difficult to break without permanent emotional damage to the bird.

Apart from the Chinese and the African, domestic breeds of goose derive from the

wild European Graylag geese. There are numerous breeds of little commercial importance to be seen from time to time in showrooms and in small backyard flocks. These include the Pomeranian, Buff, Egyptian, and the Sebastopol. The latter has curled feathers giving it an unkempt appearance and rendering the bird less waterproof than other geese.

Turkeys

Turkeys are native to North America. Early European explorers returning home with some of these birds thought they were related to the guineafowl. The name turkey may have come from the frequent "turk-turk-turk" call which is made by both sexes as they go about their normal activities. It is argued, on the other hand, that the name derives from *toka*, an Indian word for "peacock" which became *tukki* in Spanish and later "turkey" in English. The turkey adjusted to life under man's control in Europe where it was truly domesticated and was later returned to its native land as a domestic fowl.

The baby turkey, called a poult, is capable of extremely rapid growth. With a day-old weight of only 1.75–2.10oz (50–60g), some male turkeys may weigh over 30lb (13.5kg) at 25 weeks and approach 50lb (22.5kg) by one year of age.

This rapid growth requires a well-balanced diet high in protein. The protein requirement is further boosted by the rapid growth of feathers which are largely made of protein. Levels of 28–32 percent protein are used to assure rapid initial body and feather growth.

Commercially the turkey is a meat producer. Since birds with white feathers produce a cleaner-looking carcass than their dark-colored cousins, the commercial emphasis has mainly been on the white-feathered birds. It is estimated that in 1983 over 80 percent of the turkeys produced in the world had white feathers, even though the wild turkey was of a bronze color with a slight show of gray and white.

Several color variations have been selected. These were the initial basis for dividing turkeys into breeds. The colors range from solid black to totally white. Between these at least six distinct combinations have been established: Bronze, a metalic sheen of gray-brown giving a copper to gold reflection when viewed in the sunlight; Slate, an even shade of grayish blue sometimes noted with tiny black flecks; Narraganset, black with each feather tipped in white; Buff, about the shade of wheat grains; Bourbon Red, an even shade of medium dark chestnut red with white main wing and tail feathers; Royal Palm, white with black banding on the extremity of each feather—varying to give a nearly black appearance to the back.

▲ **Breeds of ducks.** (1) Aylesbury (England: meat). (2) Mandarin (China: ornamental). (3) Indian Runner (India: layer). (4) Muscovy (South America: meat). (5) Pekin (China: meat). (6) Khaki Campbell (Britain: layer). (7) Rouen (France: meat). (8) Black East India (England: ornamental).

▶ **Breeds of geese and turkeys.** ABOVE (1) Brecon Buff Goose (Wales: ornamental). (2) Norfolk Black Turkey (England: ornamental). (3) Toulouse Goose (France: meat and down). (4) Emden Goose (Germany: meat). (5) Buff Turkey (USA: ornamental). (6) Beltsville White Turkey (USA: meat). (7) Chinese Goose (China: meat).

▶ **One of the dominant breeds of goose,** the Toulouse has been selected for size and rapid early growth, but low fertility and dark feathers have made it commercially less important than the all-white Emden. Both breeds derive from the European Graylag. Two other widespread breeds, the Chinese and the African, are domesticated forms of the oriental Swan goose.

A selection for a heavily muscled carcass has led to the designation "broad breasted" for modern commercial strains. The Bronze, White Holland and Beltsville have been selected for this extreme muscle development in the breast area. The Beltsville Small White, a two-thirds-sized version of the White Holland, was developed because reduced size was felt to have increased appeal to small families.

As breeders continued to select for size and meat yield, they succeeded in changing the physical conformation of the bird's body to the extent that by 1973 more than three-quarters of the world's domestic turkey production was of necessity being mated by artificial insemination. In many instances, the extremely broad breast and relatively short legs of the commercially desirable bird made it impossible to accomplish the mating act. Strains which have not received the attention of commercial breeders still have about the same body conformation as their wild ancestors.   JLS

Gamebirds

Chickens, ducks, geese and turkeys are man's most productive domestic birds but several wild gamebirds are also farmed, for a variety of purposes including food, ornament and sport.

The **Common quail** is commercially the most important gamebird. Roast quail or boiled quail eggs are table delicacies commanding high prices. Formerly, this demand was supplied by hunters and trappers and even today quail are caught on migration in India often by the use of tethered decoys. However, there is also an intensive agricultural industry growing quail for their eggs, whose roots go back to early domestications of the species in Japan and Italy. The birds are kept in tiered wire cages in groups of 40 hens and 20 cocks; their eggs fall through the mesh and are harvested for boiling or for sale fresh.

Quail have been kept for other reasons also; in ancient times, they were pitted against each other like gamecocks and the sport attracted gamblers, as Shakespeare reminds us in *Antony and Cleopatra*. Today quail are used in research too; small and easy to rear, they are useful subjects for the study of bird physiology and behavior in the laboratory.

One of the first birds to be domesticated was the **Blue** or **Indian peafowl**. We know that captive peafowl were traded in the Near East in the time of Solomon (about 1000BC) and were probably decorating Indian temples much earlier than that. Peafowl, respected by Hindus as sacred, easily become bold and learn to rely on man for food, especially around temples.

Peafowl first appeared in Europe some time before 400BC, when a certain Demos of Athens kept them in a private zoo and charged curious visitors who came from far to look at these strange birds. As in India, their spectacular, eye-covered train invested them with a symbolic and religious significance: they were adopted as birds of the goddess Hera. From here they entered Byzantine art as images of immortality, and throughout the Middle Ages the peacock was a common symbol in art and legend. Yet peafowl have never lived wild in Europe; they thrived, and still do, in a state of semi-liberty, dependent on man for food and some protection while breeding, but living at large in parks or gardens. No attempt was made at selective breeding, and so there are no special domestic varieties. Although popular as food with the Romans and at medieval banquets, peafowl were later displaced by turkeys as the bird usually eaten at feasts.

Another bird to be domesticated in ancient times is the **Helmeted guineafowl**. It was probably brought originally from North Africa, where it is now extinct, having died out in Morocco only in recent years. The bird is still found throughout Africa south of the Sahara. Plump, tasty and gregarious,

Farming Gamebirds

Compared with traditional farmyard fowl, the economic importance of gamebirds is small. Many species are hunted, but only two are reared in large numbers: the Ring-necked pheasant and the Common quail. Neither is truly domesticated.

A native of the marshes, oases and woods of Central and eastern Asia, the Ring-necked pheasant has been successfully introduced to agricultural and woodland habitats in many parts of the world, including Europe, North America and New Zealand. In order to increase numbers, gamekeepers destroy the pheasants' predators and provide winter food. Pheasants are raised almost entirely for release into the wild at about six weeks of age, to be hunted during the next winter. Released birds augment established wild populations of introduced birds. In many places the numbers involved are considerable. In the United States some 2.5 million pheasants are released each year, and hunters kill nearly 10 million, including wild birds. By far the most important pheasant-rearing country is Britain, where about 12 million birds are released and nearly 11 million shot each year, making pheasants second only to fish as a source of wild meat.

Since it is reckoned that in Britain, at the time of writing, to rear and release one pheasant costs about £11 ($14.50US), this makes pheasant-rearing a £100-million ($130-million) business. The sale of shot birds does not cover this—they fetch only

about £2 ($2.60) apiece—but some operators make profits from renting shooting rights.

Pheasant hunting has important side effects in some areas. Its pursuit has encouraged landowners on the one hand to plant and preserve thickets, woods and hedgerows, but on the other hand to kill carnivorous mammals and birds of prey.

Some partridges (mostly Red-legged and Chukhor, but also Gray) are also reared for sport in the United States, Britain and Europe. In North America the Bobwhite quail is also reared. Common quail have long been used to lay eggs for human consumption, notably in Japan, Italy and Brazil. The Brazilians are the largest producers—about 8 million eggs a year. Large numbers of adult quail are exported to the United States and Canada, where about 6 million are consumed as table delicacies every year. Eggs are not eaten widely in North America but are increasingly consumed in Europe, where quail-rearing businesses are expanding rapidly.   MWR

REARED AS QUARRY

▲ **Red-legged partridge** with brood that will eventually be released into the wild to be shot by European sportsmen. The Gray partridge has also been used, but less successfully. In the United States the Chukhar partridge is more commonly reared for this purpose. No attempts have been made at domestication of the Red-legged partridge as all of the qualities of the wild bird are desired, but hybrids of Red-legged and either Rock or Chukhar partridges are reared for the table in some European countries.

◄ **Bobwhite quail** ABOVE is a recent addition to the species reared for sport shooting. Numbers remain small. A close relative, the Common quail, was domesticated hundreds of years ago in China or Japan as a songbird and has been developed in this century for egg and meat production. Japan farms about six million Common quail, but Brazil is the major exporter of quail and quail eggs.

◄ **The Common or Ring-necked pheasant** BELOW is the most common bird reared for hunting.

guineafowl take to captivity as readily as junglefowl, and might well, in a different world, have usurped the latter's role. But although kept for food by the Greeks and the Romans, domestic guineafowl died out in the Dark Ages to be reinstated only recently. They are rarely eaten, but are kept for ornamental reasons or as watchdogs. A flock of guinea fowl will chorus loudly if disturbed at night, so gamekeepers sometimes keep them to warn of poachers in pursuit of their pheasants.

Despite many centuries as an important source of food, the **Ring-necked** or **Common pheasant** has never really become established as a domestic bird. Numerous races of Ring-necked pheasant occur in the wild throughout much of Asia, and the forms introduced to Europe and North America are a mixture of races. About AD100 the Romans developed techniques for rearing pheasants for the table when the supply of wild birds from the Caucasus began to fail, but they are today reared mainly for a different purpose: release into the wild to be shot as game.

More than 20 million pheasants are released each year into the wild (usually at the age of about six weeks) for hunters in Europe and North America, together with smaller numbers of **Red-legged** and **Gray partridges** and **Bobwhite quail**.

In recent years many of the more exotic species of wild pheasants have been imported from their native habitats in Asia and bred in captivity. One of the commonest and most spectacular is the **Golden pheasant**, now a familiar ornament in many parks and gardens, but once a fabled rarity from the hills of western China.

There is a serious side to the breeding of exotic pheasants. Many are very rare in the wild, threatened by hunting or the destruction of habitat. By breeding them in captivity, conservationists hope to save them from extinction.

One species, **Edwards' pheasant**, survives only in captivity. In another, the **cheer**, captive-bred birds have been reintroduced to their former range in the wild to try to re-establish the species in the hills of Pakistan. In this way the hobby of aviculture, allied to the techniques of commercial pheasant rearing, contributes to the rescue of endangered species.

Aviculturalists have now turned their attention to the species which have in the past proved difficult to breed in captivity, for instance the grouse or some of the cold-sensitive tropical quails. Perhaps surprisingly, in view of the interest now being shown in new domestications of wild hoofed mammals, there is no effort being made by breeders of fowl to challenge the supremacy of the chicken as man's main domestic bird. It is simply too good at its job.   MWR

A SELECTION OF POULTRY BREEDS

Abbreviation : wt: weight.

Meat and Utility Breeds

New Hampshire Red

Farmed and originated: USA.
Developed in state of New Hampshire
from Rhode Island Red. Plumage:
light red to brown with light-colored
quills; fast feathering. Features:
vigorous; fast growth; deep, plump,
curvy body; skin yellow; single comb.
wt: cock 9lb, hen 7lb. Eggs: brown.
Originally a dual-purpose breed,
became standard for American broiler
industry 1948–52.

Plymouth Rock

Farmed and originated: USA.
Plumage: several varieties; fast early
feather growth. Features: medium to
large compact body; back wide; skin
yellow, single comb. wt: cock 9–10lb,
hen 7–8lb. Eggs: medium to light
brown. Selected for rapid weight gain
and vigor. Used in crossbreeding to
provide female parent line for more
than half of meat chickens raised in
world. Originally a dual purpose
breed used for both meat and eggs.

Orpington

Farmed and originated: England.
Developed mid to late 1800s.
Plumage: several color varieties,
especially buff; feathers held loose
from body to give massive
appearance. Features: skin white;
single comb. wt: cock 9–10lb, hen
8–9lb. Heavy roaster; fair layer with
brown eggs; sometimes used in capon
production.

Sussex

Farmed: western Europe. Originated:
England. Plumage: 3 varieties—white
with black markings most popular.
Features: rectangular body, skin
white, single comb, attractive
appearance, lively forager with good
utility characteristics. Medium sized:
wt: cock 8–9lb, hen 6–7lb. Eggs:
brown. Breeding base for much of
western European poultry meat
industry. Often considered a good
layer; used in some egg production
crosses.

Dorking

Farmed: England. Probably
introduced by the Romans.
Plumage: several varieties. Features:
legs short; unique 5th toe; skin
white; carries rather long body
horizontally. wt: cock 8–9lb, hen
7–8lb. Gained enviable reputation as
meat producer in 1880s. Largely
superseded by hybrids today.

Cornish

Developed in Cornwall, England, from
Indian Game Bird; further developed
in USA. Plumage: several color
varieties: slow feathers growth;
feathers held tightly against body
giving deceptively small appearance.
Features: extremely well-muscled
body, fiercely competitive; pea comb.
wt: cock 9.4–10.5lb, hen 8–9lb.
Eggs: brown. Matures slowly in pure
form but very useful in crossbreeding.
Today mainly an exhibition fowl.

Wyandotte

Farmed and originated: USA. Derived
from Dark Brahmas and Spangled
Hamburgs. Plumage: many varieties.
Features: wide, deep, curvy body
carrying good covering of meat on
medium-sized frame; rose comb; skin
yellow; stands cold weather well: wt:
cock 8–9lb, hen 6–6.6lb. Eggs:
medium brown. Developed as a dual-
purpose fowl, but has good reputation
as meat producer in home and
general farm flocks.

Egg-laying Breeds

Leghorn

Farmed: worldwide. Originated:
Mediterranean region; named for
Italian city of Leghorn. Plumage:
many color varieties; tightly held
feathers. Features: large comb and
wattles; comb both rose and single in
virtually any color; spritely
disposition. Small: wt: cock 5.5–7lb,
hen 3.7–4.4lb. Eggs: chalk white;
large for body size. Forms the base
stock for more than 75% of the
world's commercial egg production
strains.

Rhode Island Red

Farmed and originated: USA.
Developed in eastern USA and named
for state of Rhode Island. Plumage:
deep red or mahogany. Features:
rectangular body; noted for vigor and
longevity. wt: cock 8.3–9.4lb, hen
6.6–7.7lb. Eggs: brown. Not as
efficient as Leghorn in converting feed
to eggs but may handle poor housing
and poor nutrition better. Considered
a good dual purpose breed, winner of
many early egg laying contests.

◄ **Every feather tipped in
white,** a Narraganset turkey
moves with pomp through a
gladed wood. Of no commercial
importance, this elegant bird is
bred purely for exhibition. The
white markings on black
become progressively more
prominent away from the head
and towards the tail.

Australorp

Originated: Australia; developed from Black Orpington. Plumage: black; more closely held than Orpington, but similar in shape. wt: cock 7.7–8.8lb, hen 6–6.6lb. Eggs: beige to brown. At one time a world egg-laying champion: much used in crossbreeding in 1930s. 1940s: combinations with White Leghorn, called Austra-Whites, now displaced by smaller hybrids.

Barnvelder

Originated: Netherlands. Plumage: 2 color varieties—Partridge and Black. Features: single comb; deep, compact body of medium size—wt: cock 7–8lb, hen 5.7–7lb. Eggs: dark brown. Dual purpose breed with limited popularity as layer and hampered for commercial meat production by dark feathers. A popular farm flock bird in Europe.

Welsummer

Originated: Netherlands. Plumage: dark red and black in males, brown in females; closely held feathers. Features: trim appearance; good forager. Medium-sized: wt: cock 5.5–6.6lb, hen 4.4–5.5lb. Eggs: large, dark brown. Dual purpose breed.

Campine

Originated: Belgium. Plumage: 2 varieties, Golden and Silver; closely held; curved tail feathers of male minimized, so sometimes termed "hen-feathered." wt: cock 5–6lb, hen 3.7–4.4lb. Flighty. Eggs: white. Produces good numbers of eggs but often too small to market at top prices. Today Campines are raised mainly for exhibition.

Hamburg

Originated: Netherlands and Germany. Plumage: many color varieties; long flowing tail. Features: rose comb, very large white ear-lobes. Small. wt: cock 3.7–4.4lb, hen 3.3–4lb: graceful tending to be very wild. Eggs: white. Eggs often rather small although produced in good numbers. An attractive breed widely kept for exhibition purposes.

Andalusian

Originated: Andalusia, Spain; further developed in England and USA. Plumage: mainly blue; blue parents often producing some black or nearly white offspring. Features: single comb; skin white. wt: cock 6.6–7lb, hen 5–5.5lb. Eggs: white; numbers moderate. Bred in limited numbers; selected for exhibition.

Ancona

Originated: Italy. Plumage: mottled black with white tip on about 1 feather in 5. wt: cock 5–6lb, hen 5–5.5lb. Rose-combed and single-combed varieties. Eggs: white; good numbers, but often small in size. Not selected for egg size or persistence of production, so not receiving much attention among commercial producers.

Houdan

Originated: France. Plumage: white (rare) and mottled varieties with crests, beards, muffs. Features: extra toe each foot; white skin. Medium size: wt: cock 6–6.8lb, hen 4.8–5.9lb. Eggs: white: good numbers. Exhibition fowl, but with commercial potential, at one time highly regarded for its fine-grained flesh. Very attractive when day old, crests giving a "capped" appearance and the 5th toe causing a skipping gait.

Ornamental Breeds

Malay

Originated: southeast Asia, with considerable development in England early 1800s. Plumage: several color varieties; short feathers held tightly against body; drooping tail. Features: long legged, upstanding; small strawberry-shaped comb; skin yellow; overhanging brows giving both males and females a cruel expression. wt: cock often over 9.9lb. Eggs: brown, few in number.

Polish

Originated: Holland and Germany, probably first from Padua in Italy. Plumage: several color varieties including White Crested Black; bearded varieties with puff throat and face feathers. Features: easily startled because of large crests restricting vision, normally accompanied by wattles in non-bearded varieties. Small. wt: cock 5–5.9lb, hen 3.9–5lb. Eggs: white. Rarely broody. Unique appearance gives them great appeal in the showroom.

Yokohama

Originated: Asia; developed in Japan and other oriental countries. Plumage: several colors; in cock curved sickle, side hanging and saddle feathers dragging on ground; tail long; longest feathers have been known to approach 3ft in length on very old males; tail feathers not shed but continue to grow throughout the life of the male (in Japan and other oriental countries males often provided with special caretakers to help preserve their long tail feathers). Small: wt: cock 3.7–4.6lb, hen 2.8–3.5lb. Sprightly. Eggs: white; small in size and number. Strictly an exhibition fowl.

Langshan

Originated: China; developed in England and USA. Plumage: white or black; minimal on shanks and toes; long tail held at same angle as neck giving vase-like appearance in side view. Features: long-legged, upstanding; skin white. wt: 9–11lb. Eggs: dark brown. Nicknamed "the Lordly Bird."

Brahma

Originated: China; developed in USA. Plumage: 3 color varieties—dark, light and buff; feathered shanks and toes. Features: large, wt: often 13lb or more; but slow development (2 years); pea comb; skin yellow: good mothering; tolerates cold weather very well. Eggs: brown.

Modern Game

Originated: Britain. Plumage: many color varieties; feathers held tightly against body, short in both sexes. Features: long-legged, long-necked; cocks should have combs and wattles dubbed (cosmetically removed). More popular in bantam form (0.8–1.1lb) than standard size (5.5–6.5lb). Strictly an exhibition fowl, requiring some protection in cold climates.

Old English Game

Derived from fighting cocks introduced to Britain by Romans; probably of Phoenecian origin. Plumage: many color variations. Features: pugnacious disposition; cocks must be dubbed. Small: wt: 4.4–6.6lb; pit varieties with more compact and muscled body. Bantams often weighing as little as 0.8–1.1lb when mature. This breed has been changed very little in the last 500 years and probably much longer. Very active and alert, often wins or places well up in rooster crowing contests.

Bantam Breeds

Cochin Bantam
Cochin or Pekin Bantam

Originated: Asia. Plumage: several varieties, including Barred, Birchen, Black, Blue, Brown, Buff, Columbian, Golden Laced, Mottled, Partridge, Red, Silver Laced, White; soft fluffy feathers giving rounded contours. Features: small body with deceptively large appearance due to plumage; gentle and devoted mothering; docile, standing extreme confinement; mediocre layer. Normal-sized (non-bantam) Cochin with wt: 7.7–9.9lb and huge appearance due to plumage. Strictly an ornamental fowl, often used for foster mothering of pheasants and exotic game birds.

Silkie Bantam

Originated: Asia, supposedly taken to Europe by Marco Polo. Plumage: Black or White Bearded or Non-Bearded varieties; feathers hair-like, each filament hanging independently without locking into a vane, absorbing rather than shedding water and hence making bird vulnerable to damp; feet and shanks feathered. Features: crested; 5 toes; blue ear lobes; black eyes and skin; nearly black flesh; cock 50% larger than hen, which is sought after for foster mothering.

Sebright Bantam

Originated: England, developed by Sir John Sebright in mid-19th century. Plumage: Silver Laced and Golden Laced varieties; cock "hen feathered"—with same plumage as female. Features: small, sprightly, with rose comb; stands with breast pushed forward and tail spread; fairly good layer; eggs near white; chicks delicate in early stages; hen often not broody.

Japanese Bantam
Japanese Bantam or Cebo

Originated: Japan. Plumage: several varieties, including Black, Black-tailed Buff, Black-tailed White, Gray, Mottled, White; large vertical tails. Features: extreme shortening of legs, making bird appear to slide along; good layer, eggs near white; very alert, tending to be noisy. Gene producing short legs lethal in its double (homozygous) state, resulting in 25% of embryos failing to hatch; 25% of chicks that hatch have normal legs even though both parents short-legged.

Duck Breeds

Pekin
Most popular breed of duck worldwide. Originated: China. Plumage: creamy white throughout. Skin: yellow. Body: long, rectangular, carried higher in front (about 30° above horizontal); wt: 8.8–10lb. Eggs: white. Capable of very rapid growth, weighing as much as 7lb at 8 weeks; the basis of the commercial meat industry. Relatively good layer.

Khaki Cambell
Farmed: worldwide. Plumage: brown to tan. Skin: yellow. Body: rectangular, carried higher in front; medium-sized. Eggs: pearly white, medium-sized. Probably the world's best layer of duck's eggs, more than 1 per day. Active.

Rouen
Originated: France; takes its name from the city of Rouen. Plumage: immature feathers dark-colored like wild Mallard's, giving carcass a dirty appearance—a commercial disadvantage. Skin: yellow. Body: large, deep—mature wt often over 10lb; carried horizontally, keel parallel to and touching ground for entire length.

Indian Runner
Widely kept in small flocks. Originated: India. Plumage: several color varieties. Skin: yellow. Body: standing very straight, often compared to coke bottle; small—wt 3–5lb. Good layer. Eggs: nearly white; often 30oz or more per dozen. Active and noisy.

Black East India
Originated: probably England. Plumage: black with light green sheen. Body: slender, carried horizontally: mature wt: 3lb. Mainly a fancy or ornamental breed and quite rare.

Aylesbury
Originated: Buckinghamshire, England. Plumage: white. Skin: nearly white. Body: large, deep—wt: drake 7.7–9lb, female 7.2–8.3lb; carried horizontally. Other features: bill long and pinkish white, feet very light yellow. Raised commercially for limited meat production; also a fancy breed.

Cayuga
Originated: USA. Plumage: black and glossy. Skin: yellow. Body: rather compact, carried very slightly above horizontal; medium-sized—wt: drake 7.4–8.3lb, female 6.6–7.4lb. Mainly kept for exhibition. Has a following among home producers, but black feathers make carcass poor commercial contender; quite rare.

Call
Originated: USA and England; developed simultaneously in both countries. Plumage: several color varieties. Body: compact, oval, carried horizontally; small—wt:1.1–2lb. Other features: large, round head; prominent eyes; female exercises her loud voice frequently. Developed originally as a live decoy to call in wild ducks for hunters; has considerable following as ornamental and exhibition fowl.

Muscovy
Muscovy or Turkey duck
Originated: S America. Plumage: several color varieties, from nearly all black (wild type) to pure white; frequently slate blue and chocolate; drake lacks reverse curled tail feathers characteristic of other male ducks and has a few longer feathers on back of head which can be elevated to resemble a crest. Skin: light yellow; exposed, rough red patches on sides of head. Body: wide, flat, carried horizontally; males appreciably larger than females—wt: drake 11.6–12.7lb, female 6.6–7.7lb. Other features: hens good mothers, capable of limited flight—will perch on branches; males sometimes aggressive. Not a true duck, but behaves more like duck than like any other fowl. Eggs incubate 35 days—7 days longer than other ducks; crosses with other ducks—"mule ducks"—usually sterile.

Goose Breeds

Toulouse
Widely distributed in small numbers in N America and Europe. Originated: France, in the vicinity of the city of the same name. Plumage: medium gray on back, wings, upper body, neck and head with each feather edged in a lighter shade; lighter gray on throat, lower breast and underbody; feathers loosely held. Skin: light yellow. Body: horizontally held; wt: up to 30lb. Egg production often as little as 10–15 per annum; fertility often 50% or less.

Other features: dewlap; very short legs making bird appear to slide along ground, giving very awkward gait; very rapid initial growth but full development taking 2 or more years. Smaller, more agile variation, sometimes called "commercial Toulouse." With better fertility, reaches market wt of 14.3–15lb in 18–20 weeks; much more widely distributed in USA and central Europe.

Emden
Farmed: worldwide. Originated: Emden, Germany. Plumage: pure white; selected for early feathering. Skin: yellow. Body: long, deep—wt: 20–25lb. Egg production can be 60 or more per annum. Other features: bright yellow-orange bill and feet; rapid growth. The basis of most of the commercial goose industry.

Chinese
Widely kept. Originated: China. Plumage: Brown and White varieties; feathers tightly held. Skin: yellow. Body: compact, carried higher in front: small—wt: 10lb. Egg production: best of any geese. Other features: long, slim swan-like neck; hard, skin-covered, globe-shaped outcropping at junction of upper bill and head; noisy, active, agile. Serves well as weeder. Frequently crossed with Emden to produce market geese, but carcasses may be underweight. Used to improve egg production of other breeds.

African
Widely distributed in USA and Europe but found mainly in small flocks. Originated: probably China, despite its name. Plumage: brown and tan. Skin: light yellow. Body: deep, carried almost horizontally; large—wt:21–25lb. Other features: loud, penetrating voice; knob on upper bill like the Chinese; sometimes prone to be quite aggressive. A good commercial meat producer; ornamental fowl.

Turkey Breeds

White Holland
Farmed: most temperate regions. Originated: Europe, mainly England; a mutation of the Bronze. Plumage: pure white. Body: broad breasted; wt: tom 28–33lb, hen 17–22lb. Other features: rapid growth. rapid feather development giving protection to the body. The basis for today's commercial turkey industry worldwide.

Beltsville Small White
Originated: United States Department of Agriculture Research Station at Beltsville, Maryland. Plumage: pure white. Body: broad breasted; wt: tom 13–17lb, female 8.8–11lb—males appreciably larger than females. Developed and promoted originally as a family-sized turkey, its appeal has declined as turkey parts became available to consumers.

Bronze
Originated: in Europe from wild stocks brought from the USA and Mexico, first to Spain and later to England. Plumage: gray-brown with metalic sheen. Body: broad breasted; wt: tom 28–33lb, hen 17–22lb. The Bronze provided the first standard for the industry. Selection was made for broader more meaty carcasses, and eventually the white feathers of the White Holland were transferred to the Bronze body to make up today's commercial turkey. The wild turkey with its bronze color pattern still exists in many parts of North America and is making a comeback as a game bird.

Buff
Originated: New Jersey, USA. Plumage: a light, even shade of buff, probably selected from a mutation that occurred in the Bronze. wt: tom 22–24lb, hen 17–20lb. Rather rare; only a few bred for exhibition.

Bourbon Red
Originated: USA. Plumage: an even shade of dark red throughout except for white primary and secondary wing feathers and main tail feathers. wt: tom 22–24lb, hen 17–20lb. At one time quite popular as a commercial bird but failure to breed for broad breast has caused it to lose favor. Now bred mainly in small flocks for exhibition only.

Narraganset
Originated: USA. Plumage: black, each feather tipped with white, the white markings becoming larger towards the tail. wt: tom 22–24lb, hen 17–20lb. A beautiful bird today kept only for exhibition.

Black
Originated: USA. Plumage: solid black throughout. wt: tom 22–24lb, hen 17–20lb. Quite rare—kept for exhibition only.

The Egg

Diverting nature to human ends

The egg is the hen's half of the reproductive process. Its development as a commercial product, involving striking changes in the hen's management, is a fascinating example of how man can divert nature to his own purposes.

A female hatches with upwards of 3,000 cells in her ovaries, each with the potential of becoming an egg. As she approaches maturity her hormonal system causes them to begin to develop one after the other. In wild chickens this happens in the spring.

The ovum, the female reproductive cell, is attached to a yolk which is grasped by the funnel-shaped end of the oviduct (the infundibulum). Here the ova may be fertilized by a sperm, which has traveled an average of 22in (55cm) from the cloacal end of the oviduct where the male has deposited it. Fertilized or not, the yolk moves through the tortuously winding oviduct receiving, layer upon layer, its white and its shell. About 25 hours after the process begins, the hen lays a completed egg and within a few minutes the next yolk is released into the top end of the oviduct. After three or four days the hen will hold an egg over and lay again the following morning.

When she has 10–15 eggs a wild hen becomes broody—she stops laying and enters a state of semi-torpor in which body temperature drops about 3 Fahrenheit degrees (1.5 Celcius degrees) and the metabolic process slows. This allows her to remain in one position on the nest for long periods without muscle cramp or fecal discharge, providing warmth to the eggs.

The broody hen shifts herself on the nest from time to time, and this gradually rotates the eggs, preventing the embryos from adhering to their shells. By walking through dew-laden grass early in the morning, she provides humidity, to prevent dehydration of the eggs' contents. The movement of air through her feathers ventilates the eggs, carrying away the carbon dioxide given off by the developing embryos and replenishing the supply of oxygen.

If her nest is destroyed this broodiness will cease. If it is still early in the season, the hen may lay again in a second attempt to reproduce. When the season has advanced too far, she will not nest a second time, because there is no point in bringing on a brood when the young will not have enough time to mature before winter.

It is the hen's sensitivity to the length of the day that triggers her springtime egg laying. As soon as men understood this, they began using artificial light to give laying

▶ **Industrial egg production** ABOVE begins by breeding efficient hybrids. The hatchery may sell chicks to independent farmers or be part of a single, large-scale farming enterprise growing its own feed and distributing its own produce.

▶ **Formation of an egg.** MIDDLE The laying hen's reproductive tract, showing the oviduct from infundibulum to cloacal opening. BELOW A completed egg in cross section. When a hen is laying, 3–4 ova mature and rupture from follicles in her ovaries one after the other during intervals of 4–5 days. Each is attached to a yolk before being grasped by the infundibulum, where it may be fertilized. Layers of albumen, membrane and shell are deposited over the yolk as it travels slowly down the oviduct during a period of about 25 hours.

▼ **Hatching out.** Hens once occupied with brooding the next generation are now reserved for food production. The next generation, whether of layers or of meat birds, is a matter for commercial hatcheries, where four-line hybrids are made to industrial specifications.

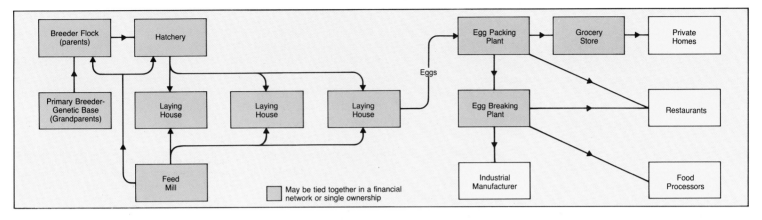

May be tied together in a financial
network or single ownership

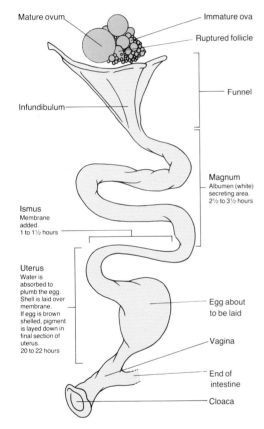

Mature ovum

Immature ova

Ruptured follicle

Infundibulum

Funnel

Ismus
Membrane
added.
1 to 1½ hours

Magnum
Albumen (white)
secreting area.
2½ to 3½ hours

Uterus
Water is
absorbed to
plumb the egg.
Shell is laid over
membrane.
If egg is brown
shelled, pigment
is layed down in
final section of
uterus.
20 to 22 hours

Egg about
to be laid

Vagina

End of
intestine

Cloaca

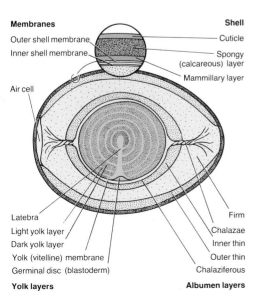

Membranes

Outer shell membrane

Inner shell membrane

Air cell

Latebra

Light yolk layer

Dark yolk layer

Yolk (vitelline) membrane

Germinal disc (blastoderm)

Yolk layers

Shell

Cuticle

Spongy
(calcareous) layer

Mammillary layer

Firm

Chalazae

Inner thin

Outer thin

Chalaziferous

Albumen layers

hens a perpetual spring. In addition, broodiness in layers has been very nearly eliminated by selective breeding. The modern egg-type hen lays about 240 eggs per year, furnishing food for human consumption that is 7.5 times her own body weight of about 4lb (1.8kg).

Hens from good strains usually lay their first eggs at 20–23 weeks of age. The first ones are often rather small, 18–21oz (510–595g) per dozen. Size increases rapidly at first and then plateaus at about 0.9–1oz per egg or 24–25oz per dozen, although a slight increase continues throughout the production period. Production may last from a few weeks for some exhibition breeds to about a year for good production birds under proper management.

Egg producers usually keep hens for 11–13 months of continuous production and then replace them with a new flock. Hens produce more eggs and consume less feed per egg laid during their first year of lay. Well-managed layers will produce a dozen eggs on 4lb (1.8kg) of feed.

The shape of the egg will vary slightly from hen to hen but not from one to the next in the same hen. Egg shape is inherited, certain strains laying shorter more ball-shaped eggs than normal, others laying more elongated ones.

In most countries there is a greater commercial demand for white-shelled eggs than brown-shelled ones. Some would argue in support of this preference that white eggs are easier to candle—a light held behind them to show interior condition will penetrate better. It is also easier to match the colors of a dozen white eggs—slight variations in brown shells show up when 12 are laid side by side. To many consumers a white shell looks cleaner.

Others would argue that a speck of dirt or a stain is more obvious on a white surface; also that a brown shell is stronger. That is usually true. The brown-shelled eggs come from the larger breeds and these are

normally, but not always, kept in flocks and allowed some range or floor pen management. Under these conditions they lay fewer eggs and can build a better shell.

The next step in the argument is to say that brown-shelled eggs must be more nutritious, since they come from birds kept in less artificial conditions than the smaller white-shelled breeds which are better suited to battery cage farming. It can be replied that nature intervenes and stops egg production if hens don't have enough of the ingredients needed to produce a "normal" egg. It cannot be denied, however, that the yolk color of an egg is affected by the hen's diet. If she is running about eating green grass, the yolk will have a darker, richer look. But this may be of no nutritional importance, only one more of the brown-shelled egg's subjective visual attractions.

The breeds of chickens developed in the British Isles are relatively large, and most lay brown-shelled eggs. Hence British people are accustomed to this color of egg and there is often consumer resistance to white shells. The same preference for brown-shelled eggs can be observed where shoppers are mostly of British or Irish descent—for example, in Boston, Massachusetts. In the Mediterranean area white-shelled eggs predominate, as they do wherever Mediterranean peoples have provided the bulk of immigrants and initial poultry stocks.

In parts of South America there is acceptance of the blue shell color of the Araucanas breed. When there is a brown pigment on the outer surface of the shell, the blue is much deeper. When Araucanas are mated with white-shelled chickens, the progeny lay eggs with a bluish cast. Mated with brown-shelled chickens, they have progeny which produce eggs with an olive or greenish cast.

Most breeds which characteristically have red ear lobes lay brown-shelled eggs. Breeds with white ear lobes tend to lay white-shelled eggs. JLS

FISH, SHELLFISH, WHALES

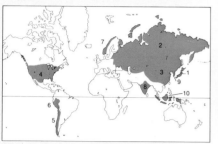

Many species from classes Agnatha, Chondrichthyes, Osteichthyes (fishes), phylum Mollusca (mollusks), class Crustacea (crustaceans) and order Cetacea (whales)

Top ten fishing nations (million tonnes caught)

1	Japan 10.7	7	Norway 2.6
2	USSR 9.6	8	India 2.4
3	China 4.6	9	Korea 2.4
4	USA 3.8	10	Indonesia 1.9
5	Chile 3.4		(FAO data 1981)
6	Peru 2.8		

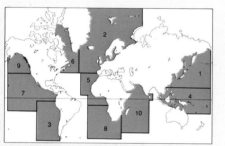

Main marine fishery areas (million tonnes caught)

1 Northwest Pacific 19.8
2 Northeast Atlantic 11.7
3 Southeast Pacific 6.9
4 Western Central Pacific 5.9
5 Eastern Central Atlantic 3.2
6 Northwest Atlantic 2.8
7 Eastern Central Pacific 2.6
8 Southeast Atlantic 2.3
9 Northeast Pacific 2.3
10 Western Indian Ocean 2.0

(FAO data 1981)

Taking fish and other animal food from the sea has always been counted a type of farming, even if only in the poetic image of the fisherman plowing his furrow through the waves or bringing in his harvest in nets. With modern management of wild stocks the comparison can only grow, as must the importance of an already ancient human enterprise—rearing aquatic creatures under genuinely controlled conditions.

Harvesting Fish, Shellfish, Whales

The ancient Egyptians and Chinese began keeping fish in ponds for ornament, and the prophet Isaiah mentions fish ponds. In 460BC, by which time fish were being farmed for food, a Chinese book on carp cultivation was written. Oyster cultivation was practiced by the Romans, and during medieval times a ready supply of fresh fish was always available from the castle moat or monastery "stew" pond.

All this time, sea mammals and fish caught in the wild provided a vastly greater portion of man's food than fish farming could. Caught fish are still overwhelmingly more important; but modern whaling and fishing efficiency have so reduced stocks, at a time of exploding world population, that aquaculture has begun to bear an increasing burden of the world's protein production. The following pages describe first marine fisheries, now nearly at a standstill in volume of produce; then areas of growth—shellfish and finned fish farming and finally whaling, a sea harvest which has virtually brought itself to an end through overexploitation.

Marine Fisheries

The seas and oceans of the world are not all equally good for fishing. The most productive areas are the shallow seas overlying continental shelves. These waters are fertilized with nutrients by run-off from the land and by mixing of tides and waves. Nutrients support the growth of minute animals and plants (plankton) in the surface waters and this starts the food chain which leads to fish.

The most productive and heavily exploited fishing regions in the world are in the northeast Atlantic and northwest Pacific which together yield nearly half the annual world catch. Both have extensive areas of continental shelf and the adjacent land masses—Europe and East Asia—are densely populated by developed nations with long fishing traditions and well-equipped fishing fleets.

Other important areas include the west coast of South America and southern Africa. In both of these places water from the deep sea, rich in nutrients, upwells at the surface generating strong plankton growth and important fisheries. Some fishing grounds, in particular the Antarctic and Indian Oceans, are underexploited because they are remote from the developed nations. The least productive waters of the world are the centers of the great oceans where there is little mixing and nutrients and plankton are scarce.

Two major groups of fish, the clupeids and the gadoids, make up one-third of the annual world catch. Clupeids include herrings, pilchards, sardines and anchovies. They are small, silvery, pelagic fish (they live in the surface waters of the sea) feeding on plankton. Gadoids, in contrast, are quite large fish; they are demersal (live close to the bottom), and feed on invertebrates such as crustaceans and on other smaller fish. They include cod, haddock, hake, pollock and whiting.

Other important surface fish include mackerel and tuna, while flat-fish and perch-like fish are important bottom-living groups. Crustaceans, including shrimps, lobsters and crabs, are important because of their high retail value, but they only account for about 5 percent of the annual world catch.

Japan and the USSR together take nearly a third of the annual world catch. This is because they have both invested in well-equipped, long distance fleets capable of fishing for extended periods anywhere in the world's oceans, from the Falkland Islands to the Seychelles to the English Channel. Other important fishing nations include those with modern medium-range fleets and good fishing grounds near at hand, such as the United States, Peru, Korea, Norway, Denmark, Thailand and Spain.

The two most important types of commercial fishing gear are the trawl and the purse seine. Trawling consists of towing a net bag, the trawl, behind a ship. The trawl is usually towed on the bottom where it catches bottom-living fish such as flat fish, cod and haddock, but when towed at high speed and suitably buoyed, it can be used in mid-water to catch schooling fish such as mackerel.

Purse seining is used at the surface to catch schooling fish such as herrings, pilchards, mackerel and some tuna. The purse seine is shot around a school and the object is to catch them all. Once the school is surrounded the bottom line is hauled in

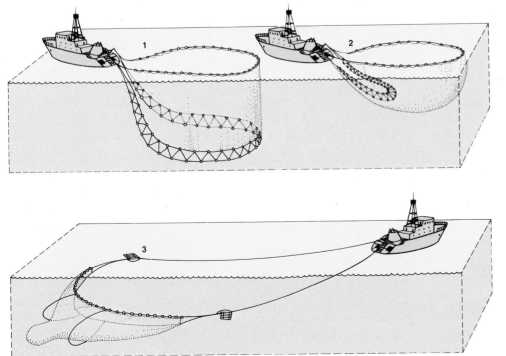

◄▲ **Seining and trawling.** Seines are used to catch fish at the surface. They may be used in shoals, as by these shore fishermen ABOVE on the east coast of Malaysia, or in deep sea commercial operations, as BELOW (1) and (2). Floats, giving the seine buoyancy, are opposed by weights which hold the seine vertical while the fishermen draw it around the catch (1). Shore fishermen may simply draw the vertically held seine back to shore. At sea, a bottom line draws the weighted ends of the net together—"purses" the seine (2)—and brings the catch back to the ship. Trawling (3) consists of towing a net bag beind a ship, either along the bottom or in mid-water.

◄ **A dying industry** OPPOSITE Fin whale on the slipway at a land-based whaling station near Reykjavik, Iceland. Members of the International Whaling Commission have agreed a pause in whaling from 1986.

World Importance of Fisheries

In terms of fresh weight the annual marine fisheries harvest is less than 1 percent of world agricultural production. However, this harvest is 95 percent animal (86 percent marine finned fish) whereas agricultural production is 94 percent plant. Thus, on a global scale, about 15 percent of animal food comes from the world's marine fisheries. Nearly 6 percent of the total global human consumption of protein comes from fish and shellfish.

The global average consumption of fish and shellfish is about 26.5lb (12kg) per person per year but the extremes range from less than 1.75oz (50g) in Afghanistan to 88lb (40kg) in Iceland. Many countries are heavily dependent on fish for protein. In Japan and some Scandinavian countries about 25 percent of all dietary protein comes from fish, and in many developing countries where animal protein is scarce, fish may form a vital component of an otherwise barely adequate diet.

Meat and fish are more expensive than foods derived from plants. Some fish and shellfish are so highly prized that they are more expensive than best-quality meat. Examples include salmon, Bluefin tuna, caviar, lobster and oysters. The global fisheries industry thus has a very great financial importance: the world catch in 1977 was worth $17 billion. Not all species of fish, however, are highly prized for direct human consumption; nearly a third of the total catch is converted to fish meal for agricultural use, mainly as an ingredient in poultry and pig feeds.

Top 20 fish

1981 catch in tons

Alaska pollack* (*Theragra chalogramma*) 4,166,550
Japanese pilchard (*Sardinops melanosticta*) 3,614,454
Chilean pilchard (*Sardinops sagax*) 2,810,605
Capelin (*Mallotus villosus*) 2,787,911
Atlantic cod* (*Gadus morhua*) 2,319,075
Chub mackerel (*Scomber japonicus*) 1,765,024
Chilean jack mackerel (*Trachurus murphyi*) 1,687,029
Peruvian anchovy (*Engraulis ringens*) 1,550,313
Pilchard (European sardine) (*Sardina pilchardus*) 1,017,252
Atlantic herring (*Clupea harengus*) 955,520
Poutassou* (Blue whiting) (*Micromesistius poutassou*) 892,179
Largehead hairtail (*trichiurus lepturus*) 764,250
Skipjack tuna (*Katsuwonus pelanis*) 697,760
European anchovy (*Engraulis encrasicolus*) 651,056
Pacific thread herring (*Opisthonema liberate*) 639,635
Cape horse mackerel (*Trachurus capensis*) 618,694
Sprat (*Clupea sprattus*) 615,014
Common mackerel (Atlantic mackerel) (*Scomber scombrus*) 613,494
Gulf menhaden (*Brevoortia patronus*) 552,567
Yellowfin Tuna (*Thunnus albacares*) 526,345

*bottom living (demersal) species, all the rest are surface dwellers (pelagic)

▲ **Important species of commercial marine fish.** (1) Atlantic cod (*Gadus morhua*). (2) Skipjack tuna (*Katsuwonus pelanis*). (3) Poutassou or Blue whiting (*Micromesistius poutassou*). (4) Capelin (*Mallotus villosus*). (5) Peruvian anchovy (*Engraulis ringens*). (6) Yellowfin tuna (*Thunnus albacares*). (7) Sprat (*Clupea sprattus*). (8) Atlantic herring (*Clupea harengus*). (9) Pilchard or European sardine (*Sardina pilchardus*). (10) Common (Atlantic) mackerel (*Scomber scombrus*).

▶ **Beloved of heartily breakfasting nations,** these adult herrings will be split, salted and smoked and sold as kippers. Immature herrings are among the fish canned and sold as sardines, although in northern Europe it is more common to pickle them. The fish swim near the surface in temperate and cooler regions of the North Atlantic and are caught by purse seining.

"pursing" the seine and trapping the fish in a bowl of netting.

To use these fishing methods fishermen need to know exactly where schools of fish are, so as to guide the net around them. Modern fishing boats are equipped with sophisticated sonar so sensitive that they can detect a single fish 1,600ft (500m) away.

Two other important methods are set-nets and long-lines. In both methods the fishing gear is shot and left to fish for several hours. Many miles of nets or lines may be set and hauled per day by a single boat. Set-nets can be hung from floats at the surface (drift nets) to catch surface fish such as herrings, pilchards, mackerel and sea-going salmon, or, more heavily weighted, they can be set on the bottom forming a wall netting and catching bottom-living fish and crustaceans. Set-nets catch fish by entanglement.

Long-lines carry hundreds of baited hooks strung out at regular intervals and in both cases the object is to catch large fish. They are set at the surface to catch tuna, sharks and sword fish, and on the bottom to catch halibut and skate.

Many stocks of commercially important fish suffer from overfishing. Fisheries scientists, by studying the catches from year to year and the growth rates of the fish, can work out the maximum sustainable yield—how much can be caught per year from a particular stock without the annual catch becoming smaller. Maximum sustainable yields can be exceeded in two ways. Slow-growing fish such as cod and plaice suffer from growth overfishing: they are given too little time to grow before they are caught and the catch decreases because the fish are all small. Faster-growing fish such as herrings, pilchards, sardines and anchovies suffer from recruitment overfishing: they are given too little time to breed before they are caught and the catch decreases because the stock is not being replaced. This is more serious than growth overfishing: it can produce the collapse of the fishery, such as happened to the Peruvian anchovy, the South African pilchard and the North Sea herring.

The annual catch of the world marine fisheries has been fairly stable for the last ten years at 60–70 million tons. It is estimated that this could be increased to 100–120 million tons by exploiting new grounds, such as the blue whiting and krill (a surface shrimp), and by conserving existing stocks. Currently overfished stocks need to be properly managed by means of international agreements on quotas, mesh sizes, nursery grounds, etc, and these agreements must be effectively enforced. The advice of fisheries scientist must be taken: a free-for-all leads inevitably to overfishing. GFW

Finned Fish Farming

Fish are reared as bait, for ornament or for release into angling waters, but almost all fish farming is for human food. In concept it is similar to conventional livestock production: stock housed in a pond, tank or cage are carefully husbanded.

Fish adopt the temperature of the water in which they live, and as its temperature increases, so their metabolic rate and food consumption increases. This means that fish grow faster when the water temperature rises, although if it gets too high, there will be insufficient oxygen. Some fish, such as trout, are adapted to living in relatively cold waters, and can even survive under ice. Tropical species require much warmer waters.

Marine fish such as tuna or turbot can be farmed only in seawater. Other marine fish such as mullet may be farmed in brackish or even fresh water. Certain fish, for example salmon, live their early life in fresh water and then migrate to the sea for a period before returning to breed. The salmon farmer needs both a fresh water hatchery and a seawater fattening unit to grow fry (young) through to marketable size.

Carp of various types are by far the commonest farmed fish in tropical and temperate climates. A considerable variety of other freshwater and marine species are also farmed, and the list is growing (eg Channel catfish, *Ictalurus puctatus*, in South America and in Mississippi and Louisiana in the United States). The ability to spawn, hatch eggs and rear larvae in captivity are essential attributes of a farmed species unless there is a ready supply of wild-caught juveniles. It is vital that the particular feed requirements of the species be easy to satisfy leading to efficient conversion of foodstuffs into fish. The species must be hardy, enabling the farmer to stock it intensively without suffering losses from stress or disease.

The wide range of fish species farmed, each with its own requirements, calls for a wide variety of husbandry systems. Even with just one species, such as Rainbow trout (see box), some farmers will specialize in egg, fry or fingerling production, others in live fish to stock "put and take" ponds for anglers, while most concentrate on rearing fish to harvest for the table market.

The greatest diversity is found in carp farming systems. In the simplest, lakes are stocked with either a single species (monoculture) or more commonly a combination of different species (polyculture), and by careful management higher yields of fish than normal occur. In the most complex systems, higher stocking densities can give better yields because aerators or flowing water maintain oxygen levels inside special fish tanks or cages provided with prepared diets. Between these two extremes lie the majority of fish farming systems, which use

Operation of a Trout Farm

The earliest attempts at trout farming were to rear Brown trout for release into angling waters. It soon became apparent that the Rainbow trout of Canada and the United States was hardier, grew faster and could be farmed commercially for the table market.

The typical Rainbow trout production cycle starts with manually squeezing the eggs and sperm (known as milt) out of female and male broodstock which have become sexually mature during the late fall. To allow fertilization the eggs and milt are simply mixed together in a bucket and the fertilized eggs are then quickly transferred to hatchery trays or incubators with a continuous circulation of clean, cool, well-oxygenated water.

The speed at which the embryo develops inside the egg depends on water temperature. After 1–2 months it hatches out as an "alevin" sustained by the food reserves within its yolk sac. After the yolk sac is exhausted,

the alevin swims up to the surface of the hatchery trough to find food. Small particles of specially prepared food are offered to the fry until first feeding is fully established. Then the fry are usually transferred to specialist concrete or fiberglass fry tanks until they are planted out into larger earth ponds, tanks or floating cages at about 0.18oz (5g).

Trout are fed compounded diets with high levels of good quality protein (eg fish meal) often delivered by automatic feed hoppers. As they grow they need to be graded into different sizes so that the population within each rearing unit is fairly uniform. Apart from feeding and grading, daily routines include cleaning pond inlet and outlet screens and harvesting market-size fish. In general the production cycle from hatched eggs to 7oz (200g) marketable fish will last from 10–18 months depending on water temperature, but many trout farmers buy in live fingerlings and have a slightly shorter cycle. CJS

▶ **Plan of a river-fed trout farm.** In the Danish trout farming system a stream is dammed and diverted into a series of up to 60 ponds 100ft (30m) long and 33–40ft (10–12m) wide. They are connected to a central channel returning to the river.

◀ **On the long way to market,** BELOW Rainbow trout in Australia swim quietly at mid-point in their progress from hatchery tray to kitchen. A native of North American streams, they perform well in fish farm conditions, reaching market size of about 7oz (200g) in 10–18 months from hatching. In their native range they are used to stock lakes for sportfishing.

▼ **Farming salmon.** Salmon at Issaquah in Washington State, USA, are carefully netted for the removal of eggs and milt, the first stage in the rearing of a new generation. The eggs are fertilized by artificial mixing with the milt. Young hatched out in fresh water troughs reach market size in saltwater fattening tanks, imitating their natural migratory life-cycle, by which adults live most of their lives in the sea before breeding in freshwater streams.

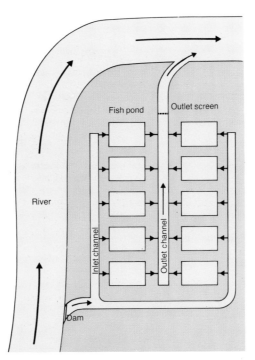

stagnant ponds, for example, enriched by manure or inorganic fertilizers in order to stimulate the growth of plant and animal life on which the fish can feed, with some supplementary food usually given directly by the farmer.

Under tropical conditions, plankton-eating fish such as mullet and milkfish grow well in fertile ponds. The growth of Common carp and tilapia is improved by adding agricultural by-products such as rice bran. In Chinese carp farms a grass carp is stocked to browse on aquatic plants while its waste products fertilize the water encouraging the growth of plankton; so a plankton-eating carp is also stocked, together with other carp species which feed on animals in the mud and on the feces from the other fish. This polyculture method involves stocking a number of different species in the same fertile pond so that each relies on a different food source to the benefit of the entire fish

stock. Yields of up to 4 tons per acre (10 tons per ha) can sometimes result.

Simple excavated fish ponds are up to 25 acres (10ha) in area and the water depth is approximately 3.3ft (1m). There is usually a deep harvest basin with an adjustable outlet pipe so that the entire pond can be virtually drained for harvesting. Ideally there should be separate ponds for spawning and hatching, nursing the young fry, fattening the larger fish for market and for holding male and female broodstock. The production cycle is determined by water temperature so that in a country such as Israel it is possible to obtain two crops each year of 21oz (600g) carp.

The trend in fish farming is towards higher yields. More intensive production means a more scientific approach to breeding, feeding and managing the fish stocks. Injecting broodstock with hormones encourages spawning and guarantees a supply of fry throughout the year. As more is learnt about the exact nutritional requirements of fry and larger fish, so it becomes possible to formulate whole balanced diets which can be fed as pellets to improve growth rates and reduce waste. Veterinary skills are now being used, with fish being vaccinated against disease. Perhaps the greatest scope for improvement lies in the application of genetics to improve fish stocks, which today are still largely undomesticated.

The list of fish species being successfully farmed is growing each year, particularly the higher-priced marine fish. For certain species it is becoming possible to stock the sea with hatchery-reared juveniles in order to increase the wild catch of these fish. The fact that fish can routinely grow through to marketable size under farm conditions means that fish markets will increasingly look to farms as a truly reliable source, free of the risks and uncertainties of hunting fish in the wild.   CJS

Shellfish Fisheries

Shellfish form about a tenth of the world fisheries harvest. Important groups include shrimps and prawns, squids, clams, scallops, crabs and lobsters.

Commercial species of shrimps and prawns comprise both surface living and bottom living forms. They are caught using trawls or purse seines. Their larger, bottom dwelling cousins, the crabs and lobsters, are usually caught in baited pots or traps set out on rocky ground in long strings, often with many pots per string. Some spiny species of crabs and lobsters are caught by entanglement in weighted set nets.

Most squids live near the surface of the sea. Shoaling species may be caught in seine nets or in mid-water trawls. Many squids, however, are caught on lines using baited or attractively colored "jigs"—multiple hooks shaped like small grapnels which are "jigged" up and down in the water.

Clams and scallops live on sandy or muddy bottoms and are mostly caught in dredges towed behind boats. The dredge often has teeth along the lower edge of its mouth to dig the clams out of the sand. In shallow water suction dredges may be used; these suck a mixture of sand, water and clams onto the deck and through a sieve which retains the clams.

Farming Shellfish

Most shellfish are sea dwellers with complicated life-histories. Many thousands of tiny, delicate, free-swimming larvae are produced by each spawning female and in the wild

◄ **Fish ponds and vegetable plots** crowd together beneath mountains on Lantau Island, Hong Kong, where every potentially productive corner finds a use. Mainland China also farms fish and leads the world in production. High yields are achieved from ponds stocked with complementary species of carp: grass eaters fertilize plankton eaten by other carp, and further species may feed directly on the feces of these.

▼ **Species of farmed fish.** (1) Milkfish (*Chanos chanos*). (2) Thick-lipped gray mullet (*Mugil cephalus*). (3) Atlantic salmon (*Salmo salar*). (4) European eel (*Anguilla anguilla*). (5) Catla, an Indian carp (*Catla catla*). (6) Tilapia (*Sarotherodon galilaeus*). (7) Ayu (*Plecoglossus altivelis*). (8) Yellowtail (*Seriola lalandi*). (9) Channel catfish (*Ictalurus punctatus*). (10) Common carp (*Cyprinus carpio*).

these swim in the plankton at the surface of the sea, feeding on other minute organisms.

If farming is to be attempted over the entire life of such a creature, the adults must be bred in captivity and special hatcheries developed. Conditions for raising larvae vary with each species and require careful control of temperature, salinity, oxygen and acidity.

Precisely the right type of live food for the particular larval stage must be provided. To avoid these difficulties the commonest method of farming is to collect the young post-larvae (seed) from the wild and put them in a managed area to grow, sometimes with extra feeding.

Oysters are cultured mainly in Japan (Pacific oysters, *Crassostrea gigas*), the USA (American oysters, *C. virginica*), and France (Portuguese oysters, *C. angulata*, and flat oysters, *Ostrea edulis*). In most cases seed is collected from wild populations. In Japan strings of scallop shells (imported from the USA) on which the larvae can settle are hung in the sea. In France special ceramic tiles are used. Some collecting is done with oyster shells in the United States, but much of the seed is supplied by hatcheries and some is imported from Japan.

Oysters are grown to marketable size in Japan by off-bottom raft culture. Strings of scallop shells bearing one-month-old seed (0.5in—12mm) are suspended on wires beneath anchored rafts. The oysters feed on natural plankton and grow to harvest in 6–18 months at a size of 2–8in (5–20cm). Annual production of whole oysters is up to 26 tons per acre (140 tons per ha).

The United States and France both practice a less efficient bottom culture, in which sea oysters are laid out on the sea bed in suitable areas and later dredged up or collected at low tide. Growth is slower than on rafts, stocking density is lower, and there is much loss to bottom-living predators such as starfish, crabs and whelks. Harvest time at a size of about 3.2in (8cm) is 2–5 years, depending on temperature. Annual production from bottom culture is 0.4–2.0 tons acre (1–5 tons per ha).

The most important farmed species of **mussels** is the European Blue mussel (*Mytilus edulis*). Spain and France are major producers, and in both countries seed is collected from the wild on loosely woven ropes hung in the sea. In Spain these ropes are then suspended from anchored rafts. In France they are wrapped around wooden posts driven into the bottoms of shallow bays at low tide.

The mussels feed on natural plankton and grown to a marketable size of 2–4in (5–10cm) in 12–18 months. Production in France is about 2 tons per acre (5 tons per ha) per year. In Spain the high stocking density of raft culture gives a phenomenally high productivity of up to 243 tons per acre (600 tons per ha) per year.

Clams and **cockles** live buried in sand or mud on the shore or in shallow water. Beds of various species are managed and fished all over the world. Farming is practiced in Southeast Asia and the United States. In Southeast Asia seed is collected by hand from the wild in places where plankton settlement is particularly dense. The seed is

▲▶ **Oyster culture.** ABOVE Oyster larvae on strings of scallop shells hang by wires beneath these rafts in Ago Bay, Japan. The developing larvae feed on naturally occurring water-borne plankton to reach adult sizes in 6–18 months. RIGHT This modified bottom culture, seen at low tide in Normandy, France, achieves a high stocking density and reduces losses to bottom-dwelling predators. A less efficient but common form of bottom culture is to lay out the oysters on the sea bed and later dredge them up.

then redistributed in other areas at lower densities and the crop is harvested, usually by hand, 6–24 months later. In the United States the quahog (*Mercenaria mercenaria*) is farmed through its entire life. Seed produced in hatcheries is scattered on the bottom in shallow water. Growth to harvest takes 5–8 years. As with oysters, bottom culture leaves clams and cockles vulnerable to predators.

Two other types of marine mollusk are farmed on a small scale, chiefly in Japan: **scallops** and **abalones**. In both cases wild seed is scarce and hatchery facilities are necessary. Scallop culture is similar to that of oysters, but abalones, being marine snails, crawl about and feed on seaweed so cannot

simply be left to grow. They are confined in enclosures and fed regularly.

The main **prawn** farming area of the world is Southeast Asia, with Japan leading the field. The most important species are Japanese prawns (*Peneus japonicus*), Giant tiger prawns (*Peneus monodon*), and Giant freshwater prawns (*Macrobrachium rosenbergii*). These prawns are omnivorous and in the wild feed largely on algae and detritus. Japanese and Giant tiger prawns live in estuaries and in the sea. Giant freshwater adults live in freshwater, but the larvae require brackish conditions. Prawns can be grown to marketable size of 4–8in (10–12cm) in 6–12 months.

The simplest way of growing Japanese and Giant tiger prawns, practiced widely in the coastal swamplands of tropical Southeast Asia, is to exploit the natural life cycle of the wild population. Breeding occurs offshore but the young move inshore to feed and grow until they, in turn, move

World Importance of Fish, Shellfish and Crustacean Farming

China has by far the largest national harvest of farmed aquatic produce—over 4 million tons annually—followed by Japan with over 1 million tons, mainly seaweed and shellfish. World production is close to 10 million tons, split into three roughly equal portions between finned fish (especially carp), shellfish (especially oysters and mussels) and seaweed. At least 80 percent of production is in Asia. It has more than doubled in 10 years to equal about 13 percent of the global wild fish catch.

The annual world harvest of mollusks is about 5 million tons. The figure for crustaceans is about 3.3 million tons. This is about 11 percent of the total world aquatic production. Only a small proportion of this 8 million tons of shelled sea food is produced by farming; most is caught by fishing including 3 million tons of squid, about 1 million tons of crabs and lobsters and half a million tons of krill. The great majority of the 1 million tons of clams and cockles and of the

1.7 million tons of prawns are also caught by fishing. The bulk of the harvest of oysters (1 million tons) is produced by farming, and so are most of the 600,000 tons of mussels, but cultured prawns account for only 5–10 percent of the world prawn harvest, and the annual world production of crayfish is only

Top 10 crustaceans
1981 catch in tons
Antarctic krill* (*Euphausia superba*) 448,266
Akami paste shrimp* (*Acetes japonicus*) 163,100
Northern prawn (*Pandalus borealis*) 100,818
Blue crab (*Callinectes sapidus*) 97,974
Brown shrimp (*Penaeus aztecus*) 67,469
Banana prawn (*Penaeus merguiensis*) 63,746
Pacific snow crabs (*Chionoecetes* species) 61,064
Kingcrabs (*Paralithodes* species) 60,693
Norway lobster (*Nephrops norvegicus*) 48,580
Common shrimp (*Crangon crangon*) 45,573

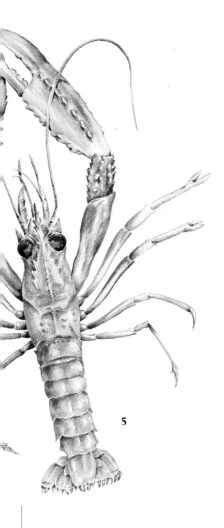

5

◄ **Important species of fished crustaceans.**
(1) Krill (*Euphausia superba*). (2) King crab
(*Paralithodes* species). (3) Common shrimp
(*Crangon crangon*). (4) Northern prawn
(*Pondulus borealis*). (5) Norway lobster
(*Nephrops norvegicus*).

Top 10 shellfish

1981 catch in tons

Pacific cupped oyster (*Crassostrea gigas*)
487,850 (mostly farmed)
Blue mussel (*Mytilus edulis*) 387,370 (mostly
farmed)
American cupped oyster (*Cassostrea virginica*)
329,437 (mostly farmed)
Japanese flying squid* (*Todarodes pacificus*)
228,355
Sea scallop (*Placopecten magellanicus*) 188,070
Japanese clam (*Venerupis japonica*) 184,857
Japanese scallop (*Pecten yessoensis*) 152,180
Calico clam (*Argopecten gibbus*) 146,773
Ocean quahog (*Arctica islandica*) 135,932
Surf clam (*Spisula solidissima*) 111,592

*Surface (pelagic) species, all the rest live on or near the bottom
(demersal).

out to breed. The prawn farmer simply constructs a pond in the tidal zone and allows the rising tide to fill it. The incoming water contains post-larval and young prawns which remain in the pond to feed and grow. Large prawns attempting to leave with the falling tide are harvested in nets. To achieve best results the ponds need to be carefully constructed (not too deep, with sufficient sluice gates) and properly managed, with supplementary feeding or fertilization and the removal of predators. This type of farming yields about 445lb per acre (500kg per ha) per year but controlled stocking with specially collected seed prawns can more than double this yield.

Intensive farming in Japan employs hatchery reared seed stocked at high densities (10 per sq ft—100 per sq m) in flowing seawater and fed entirely by the farmer on prepared foods. Annual yields of 4 tons per acre (10 tons per ha) can be obtained.

Freshwater **crayfish** look like small-clawed lobsters; they are omnivorous and have no larval stages. Several species are farmed in ponds in Europe, the United States and Australia. In the United States the main species is the Red crayfish, *Procambarus clarkii*. This is cultured by stocking shallow ponds (which may double as rice fields in alternate seasons), controlling the growth of pondweeds (some are useful as crayfish food and cover, others may have damaging effects) and controlling predators such as fish, birds and bullfrogs. The crayfish population is self-sustaining and the larger individuals are harvested by trapping or netting.

about 1,600 tons. On a global scale, compared to the annual fisheries harvest of some 70 million tons, and the land agricultural harvest of some 7 billion tons, shellfish farming fades into insignificance. But a ton of oysters or prawns is significant on a local scale, especially in Southeast Asia. Prawn tails fetched about $6.82 per pound ($15US per kilo) on the export market in 1980. The farming of these animals is of especially great importance in Southeast Asia.

It has been estimated that coastal aquaculture production could be expanded by at least ten times, but limitations on expansion include the increasing industrial pollution of coastal regions (especially of highly productive estuaries), the high levels of investment needed for intensive culture, and legal restrictions on the use of coastal waters. It can also be difficult to market unfamiliar foods and extra production can have a bad effect on prices.

Growth to marketable size of about 4in (10cm) takes 6–12 months and annual yields are up to 890lb per acre (1 ton per ha).

The high retail value of **lobsters** and **crabs** makes farming them appear attractive, but they are carnivorous, slow growing and cannibalistic when stocked at high densities. In Southeast Asia the large swimming crab, *Scylla serrata*, is often grown and harvested as a subsidiary crop, and sometimes a main crop, from coastal fish and prawn ponds. In intensively managed farms it is regarded as a pest.   GFW

Whaling

Man has hunted whales from prehistoric times until the present. Eskimos and North American Indians were whaling over 2,000 years ago, and European records date from the 9th century in Norway and Flanders.

There was an important development in the Bay of Biscay from the 10th century, where the Biscayan or North Atlantic right whale was hunted. As the stocks were reduced the whalers searched farther afield until they voyaged right across the North Atlantic to Newfoundland and the Gulf of St. Lawrence in the 15th century. The Basques, Norwegians and Icelanders were hunting around Iceland in the 16th century, no longer just for subsistence, but in a whaling industry.

As the search for whales extended, the Bowhead, Greenland or Arctic right whale was found, and this formed the basis for the northern whale fishery, centered originally on Spitsbergen in the 17th century. Basque whalemen were employed by British, German, French and particularly Dutch vessels. Later the vessels ventured along the ice-edge to Greenland and the Davis Strait in search of new concentrations of whales. Up to 150 British vessels were engaged in a year.

Americans also had an arctic fishery for the Bowhead whale, north of the Bering Straits and into the Okhotsk Sea in the second half of the 19th century, which all but eliminated the species there too. Whaling by settlers in New England started in the 17th century for right whales, but a chance catch of a Sperm whale by a shore whaler blown off course in a storm led to a major new fishery for this species in the 18th and 19th centuries. New England whalers held the monopoly of this southern whale fishery, working across the Atlantic, to the south and on into the Indian and Pacific Oceans, although the British were the first to take Sperm whales in the Pacific. In 1846 there were 729 Yankee whaleships at sea; but the discovery of petroleum in 1859 made available a competitor to whale oil and a slow decline of this fishery followed.

The worldwide hunt for Sperm whales led to the discovery and exploitation of Southern right whales around the southern continents and subantarctic islands. They were much depleted by the end of the 19th century. Shore whaling was carried on from South Africa, Australia, New Zealand and Tasmania. Whaleships and shore whalers also seriously reduced the Gray whale in the Mexican breeding lagoons and along the Californian coast.

A quite different and independent tradition of whaling had meanwhile been running parallel with these European and North American-based fisheries. The Japanese had for many centuries been heavily dependent on products from the sea for food, and they had developed a coastal whale fishery employing a netting technique to trap mainly Gray and right whales which then killed and cut up on shore.

By the mid 19th century the slower swimming whales were scarcer, and in 1864 the Norwegian Svend Foyn developed a method of hunting the fast rorquals by firing a heavy harpoon fitted with an explosive grenade from a cannon mounted on the bow of a steam vessel. This revolution in technique soon spread to many parts of the world, and the catching power it represents brought coastal stocks of the Blue, Fin and Sei whales to low levels in the North Atlantic, Arctic and North Pacific. At the beginning of the 20th century Norwegian-style whaling extended into the Antarctic, where moored factory ships flensed the whales alongside in sheltered bays or ice fields.

In 1925 the factory ship *Lancing* was fitted with a stern slipway up which whales were hauled for processing on deck. This heralded the era of open-sea whaling with expeditions comprising a factory ship and its attendant catcher vessels extending far round the Antarctic Ocean. In the 1930–31 season 41 such factory expeditions and six land stations with 232 catchers killed 40,200 whales and produced over 3.5 million barrels of oil. The whales could not withstand so great a slaughter, and the Antarctic industry has gradually declined as the Norwegian, British, Dutch, South African and German fleets dropped out, in spite of turning to successively smaller species of whales. Now only Japan and the USSR are operating, taking Minke whales which are one-tenth the size of the Blue whales which provided the early targets.

Early whaling was essentially uncontrolled and unrestricted. Modern whaling

has been subject to more regulation, usually by the licensing of the shore stations by the government concerned. Antarctic whaling was at first unrestricted on the high seas, but after the over-production of oil in the 1930–31 season the various companies made voluntary agreements to control output and hence the market price. Concern over the declining stocks then led to a series of international agreements, starting at the League of Nations in 1931 and culminating in the 1946 Convention for the Regulation of Whaling, which established the International Whaling Commission (IWC).

Most whaling nations joined the Commission and adopted agreed measures designed to conserve the whale stocks and bring about orderly development of the industry. These included minimum-size limits, open and closed seasons, protection of calves and their nursing mothers, total protection of right and Gray whales, and catch limits for the hunted species. Unfortunately, the catch limits were set too high and the stocks and industry continued to decline.

▲ **One of the last,** a Minke whale being hauled on board a Japanese catcher.

▼ **Species of whales.** (1) Fin whale (*Balaenoptera physalus*). (2) Minke whale (*B. acutorostrata*). (3) Sei whale (*B. borealis*). (4) Humpback whale (*Megaptera novaeangliae*). (5) Blue whale (*B. musculus*).

1

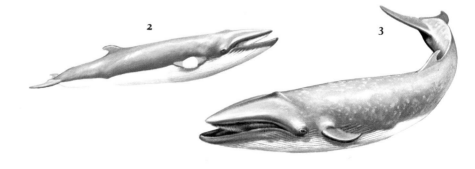

2   3

World Importance of Whaling

The early subsistence hunters caught whales for the meat and oil. The northern commercial fishery of the later centuries was concerned mainly with oil. The blubber was stripped off the dead whales in coastal bays and boiled in pots on the beaches to extract it.

Away from such shelter the whales were cut up at sea and the blubber transported back to the home port in casks. The long baleen plates were also a valuable product for corset stays and umbrella ribs.

The southern or Sperm whale fishery also concentrated on oil. Sperm oil is chemically different from that of baleen whales and superior as an oil for illumination. Sperm whales also yielded wax for candles.

In modern times the oil from baleen whales found a major market in the manufacture of margarine and high-grade cooking fat, while sperm oil had many industrial uses in quenching and rolling steel, as a dressing for leather, in lubricants, detergents and cosmetics. When whales were still abundant their oil supported a highly profitable but extremely wasteful industry.

After the Second World War, and particularly with the re-entry of Japan into the fishery, the baleen whales were more fully utilized, particularly as a source of meat. Sperm whales are also hunted far more intensively for industrial oil. During the prewar period the whale catch constituted from 12–15 percent of the weight of the total world harvest of fish, shellfish and mammals from the seas. After the war this population contributed 8–11 percent of the marine harvest until 1950 and declined to less than 1 percent by the mid 1970s.

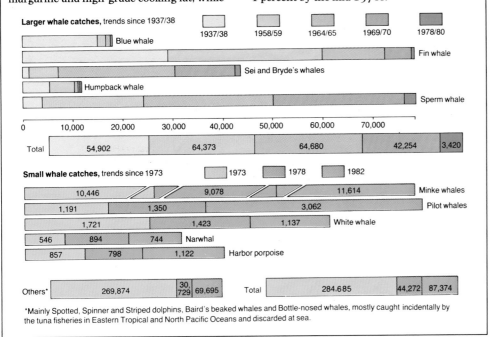

Larger whale catches, trends since 1937/38

| | 1937/38 | 1958/59 | 1964/65 | 1969/70 | 1978/80 |

Blue whale
Fin whale
Sei and Bryde's whales
Humpback whale
Sperm whale

0   10,000   20,000   30,000   40,000   50,000   60,000   70,000

| Total | 54,902 | 64,373 | 64,680 | 42,254 | 3,420 |

Small whale catches, trends since 1973

| | 1973 | 1978 | 1982 |

Minke whales	10,446	9,078	11,614
Pilot whales	1,191	1,350	3,062
White whale	1,721	1,423	1,137
Narwhal	546	894	744
Harbor porpoise	857	798	1,122

| Others* | 269,874 | 30,729 | 69,695 | | Total | 284,685 | 44,272 | 87,374 |

*Mainly Spotted, Spinner and Striped dolphins, Baird's beaked whales and Bottle-nosed whales, mostly caught incidentally by the tuna fisheries in Eastern Tropical and North Pacific Oceans and discarded at sea.

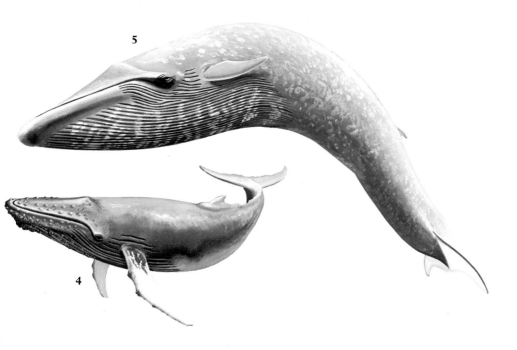

5

4

A crisis was reached in the early 1960s, and a group of independent experts in population dynamics was formed to provide a scientific basis for future regulations. From this initial work has developed the current strategy of limiting catches to what the stocks can sustain, formulated in 1974 as a new management procedure based on the concept of maximum sustainable yield.

Whales, like any other animal population, have a natural capacity for increase and a natural rate of mortality. At their initial level these two factors balance. As a stock is reduced the pregnancy rate rises, the age at which the whales start to breed falls and hence recruitment increases. At some particular stock level the surplus of recruits over natural deaths reaches a peak, the maximum sustainable yield, which can be safely harvested without reducing the stock size. At stock sizes below this, the surplus of recruits over natural mortalities declines until another equilibrium is reached at very small levels.

Catch limits for the more numerous whale stocks are set slightly below the increased reproductive potential of the whales. These stocks will therefore provide a harvest for an indefinite time. Stocks below the optimum are fully protected from commercial catching, even though they could safely provide some small yield, in order to encourage the maximum rate of rebuilding.

Unfortunately, it became clear that even though many whale stocks are among the best analyzed and assessed marine living resources, the amount of information available on stock sizes, yields and trends was inadequate to give total confidence in the harvesting strategy. Also many non-whaling nations have joined the IWC in recent years because of their interest in the potential of whales as living animals for sightseeing, educational, recreational and cultural purposes, rather than for hunting and product utilization. Together with questions over the accuracy of the stock assessments, the present 39 members of the IWC have agreed on a pause in commercial whaling from 1986, the situation to be reviewed in five years at the latest.

Whaling has gone through a cycle from coastal hunting for subsistence purposes, through a worldwide commercial industry which has boomed and declined, to an insignificant enterprise which is due to be suspended, leaving only some minor subsistence hunting in the Arctic. Catching power has exceeded reproductive capacity, and knowledge has been too little and too late to prevent over-exploitation.   RG

BEE-KEEPING

Four species of genus _Apis_
Phylum: Arthropoda.
Class: Insecta.
Order: Hymenoptera.
Family: Apidae.

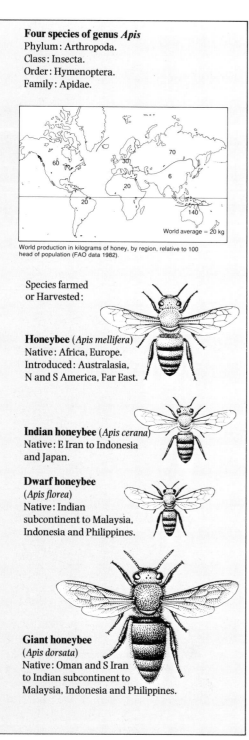

World production in kilograms of honey, by region, relative to 100 head of population (FAO data 1982).

Species farmed or Harvested:

Honeybee (_Apis mellifera_)
Native: Africa, Europe.
Introduced: Australasia,
N and S America, Far East.

Indian honeybee (_Apis cerana_)
Native: E Iran to Indonesia
and Japan.

Dwarf honeybee
(_Apis florea_)
Native: Indian
subcontinent to Malaysia,
Indonesia and Philippines.

Giant honeybee
(_Apis dorsata_)
Native: Oman and S Iran
to Indian subcontinent to
Malaysia, Indonesia and Philippines.

► **Bee-keeper with brood-frame.** Besides easy extraction of honey from reusable comb, the movable-frame hive gives bee-keepers control over the rate of reproduction in their colonies. A "queen excluder," a grid big enough for workers but not the queen to pass through, can be inserted to restrict brood space, separating the brood-frames from frames supporting the honey combs. The colony can expand without becoming crowded enough to swarm. (See diagram p143.)

Man has been keeping honeybees in hives for 4,000 years or more, but before this he was a honey hunter and robbed wild colonies of their stores and brood. Honey is still hunted in many parts of the world today; it remains the only way of gathering honey from the wild species of Asia as no means of keeping them in hives has yet been devised. Unfortunately the robbed colonies are often destroyed.

Bee-keeping methods vary enormously from the peasant who keeps his colonies in clay cylinders, pots, baskets or hollow logs, to the sophisticated bee-keeper who uses modern, accurately constructed wooden hives, makes use of the latest scientific information in his management techniques, and employs labor-saving devices.

Bees produce honey from the nectar of flowers. This is taken back to the hive and stored in the hexagonal cells of the wax combs. In more primitive types of bee-keeping, the bee-keeper cuts the honey comb from the hive and harvests the honey by squeezing it from the comb. The crushed wax comb that is left is a valuable by-product; although beeswax has been replaced by synthetic products for many traditional uses, it is still of importance in the pharmaceutical, cosmetics and armaments industries.

In the modern bee hive the combs are built in wooden frames that can be removed from the hive. The wax cappings are sliced from the honey cells, and the combs, supported in their wooden frames, are placed in a honey extractor which whirls them around at high speed throwing the honey out by centrifugal force. The empty wax combs can be returned to the colonies to be used again by the bees.

Handling and preparing honey for market has become increasingly mechanized during this century. The extracted honey is filtered, sometimes under pressure, and can be processed to produce either liquid or granulated honey, but there is now an increased consumer demand for a natural product with minimal processing.

Other products of the hive are also used. The brood food or "royal jelly" fed to queen larvae has various clinical and cosmetic uses. Pollen collected by foraging bees is marketed as a health food. Propolis, a resinous substance secreted from the buds, branches and leaves of certain trees and collected by bees to seal and strengthen parts of their nest, has antibiotic and pharmacological properties. Bee venom is used medicinally to desensitize patients who have become allergic to bee stings.

Bee-keeping throughout the world not only provides more honey but helps many crops that need insect pollination. Some crops yield little fruit (eg apples, pears), or seed (eg red clover, runner beans, alfalfa), without insect pollination. For others pollination gives only moderate increases; but adequate pollination may have other benefits such as earlier ripening of the crop and a more uniform harvest (eg field beans) or the production of fewer malformed fruits (eg strawberries).

In recent years there has been a tendency in many countries to make fields and orchards larger and to concentrate crops in particular areas. The numbers of wild pollinating insects present are no longer sufficient to pollinate them. Modern agricultural practices also tend to destroy nesting sites of beneficial insects and many of their food plants, so there is actually a decrease in the numbers of pollinating insects available. Although large areas of a single crop may provide bees and other insects with ample forage for a short time, there is little forage for the rest of the year.

This deficiency in pollinating insects has led to an increased demand by growers for honeybee colonies to perform pollination services, and many bee-keepers supplement their income by hiring their colonies for this purpose.

There are four species in the honeybee genus _Apis_. _Apis mellifera_, the honeybee of Africa and Europe, nests naturally in caves and hollow trees; it builds a series of approximately parallel wax combs which are covered on each side with hexagonal cells in which the young bees are reared and the collected food is stored. These bees were introduced by early settlers from Europe to Australia and New Zealand and to North and South America where no _Apis_ species were native, although Indians in Central and South America collected honey from colonies of stingless bees.

In 1956 strains of _Apis mellifera_ were imported from South Africa and Tanzania into Brazil: the intention was to undertake a controlled breeding program to impart greater honey productivity into the strains of _Apis mellifera_ imported some centuries earlier. Because of the well-known aggressiveness of many African honeybees the imported colonies were not allowed to swarm. However, the next year, 26 managed to escape and settle in the wild. They quickly hybridized with the local bee populations and their undesirable viciousness became established in them.

Because of the strong tendency of these

World Importance of Bee-keeping

About 50 million honeybee colonies are kept by 5 million bee-keepers throughout the world. They give an annual harvest of 600,000 tons of honey.

Methods and equipment vary greatly. Hobbyist bee-keepers, with up to ten colonies each, are predominant in Europe, which has the highest density of colonies and the lowest honey yields, about 22lb (10kg) per hive each year. In favored parts of the New World, bee-keeping operations can be highly mechanized and one bee-keeper can manage several hundred or even a thousand colonies by himself; honey yields commonly exceed 110lb (50kg) per colony and may even average 220lb (100kg) in good conditions.

In many countries, most honey is consumed locally, but there is a well-established pattern of international trade (about 230,000 tons per annum) based on the greater consumption in the more affluent countries.

Consumption is greatest in West Germany and Switzerland where it amounts to 2.2–4.4lb (1–2kg) per person per year, and in France, Italy, Japan, the United States and the United Kingdom where it approaches 1.1lb) (500g) per person. There has been a dramatic increase in Japanese consumption during the last few decades.

Because of their low productivity and high consumption, most European countries and Japan import large amounts of honey. The high production in the United States is insufficient to satisfy its own requirements, but production in China, New Zealand, Australia, South and Central America and Canada exceeds home demand and much more is exported. China has only recently become a major exporter.

Consumption, even in the affluent countries, is still not large and may well increase with interest in natural foods and health products. There should be ample opportunity for developing countries to produce and export good quality honey in the world markets once their bee-keeping has progressed sufficiently to satisfy home requirements. Countries in South America and East Africa have marked potential for developing their honey industries.

▲ **Like seasonal farm labor,** honeybees lodge in a temporary hive city on a hillside in Greece, hired out to growers of the crops in the valley bottom. Easily transported, bees are versatile pollinators.

◄ **Vulnerable nest.** With only a single comb hanging from a branch or a rock, the nest of the Dwarf honeybee (*Apis florea*) is easily destroyed when its honey is collected. This docile Asian species occurs only in the wild, as no way has yet been found of keeping it in hives.

► **Structure of a bee-keeper's hive.** One or more shallow "supers" hold honey frames which can be removed for the extraction of honey. Deep supers hold brood frames which can be removed to inspect for disease or reproductive failure. A queen excluder, a mesh small enough to stop the queen but large enough to let workers through, limits brood space and population growth. The "beespace" between combs and other structures is calculated to allow movement of the bees without being big enough to stimulate comb production where it is not wanted.

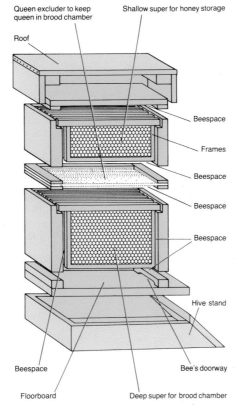

Queen excluder to keep queen in brood chamber

Shallow super for honey storage

Roof

Beespace

Frames

Beespace

Beespace

Beespace

Hive stand

Beespace

Bee's doorway

Floorboard

Deep super for brood chamber

hybridized bees to swarm and migrate, their aggressive characteristics have rapidly spread through most of South America. Colonies can sometimes be easily provoked, slight disturbances resulting in mass attacks. Although bee-keeping is still possible with such colonies and they can give good honey crops, it must be done with great care to keep the bees under control and to ensure that no other human beings or domestic animals are in the vicinity.

The Indian honeybee is a smaller bee than *A. mellifera* but has similar nesting habits, and each colony also builds a series of combs. In recent years *A. mellifera* has been introduced into the Far East and has replaced the Indian honeybee in many areas.

The other two species, the Dwarf honeybee and the Giant honeybee, occur only in the wild. They build only a single comb from a tree branch or overhanging rock, although the Dwarf honeybee also nests in cavities. The Giant honeybee is the largest and can be very aggressive. The Dwarf honeybee is the smallest and is usually quite docile.   JBF

SILK MOTHS

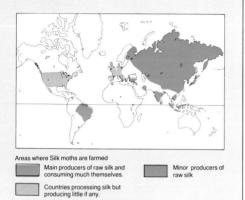

Areas where Silk moths are farmed

◼ Main producers of raw silk and consuming much themselves.

◼ Minor producers of raw silk

◼ Countries processing silk but producing little if any.

Silk Moths and Types of Silk:

Antherea assama
A partly domesticated silkmoth which yields the traditional Muga silk of Assam.

Antherea mylitta
A native moth which lives wild on oak in India. Rarely cultured but cocoons collected; hence produces, like all other species except *Bombyx mori*, a "wild" silk. This cocoon yields a strong but coarse, brown silk known as Tusset or Tasar silk. All "wild" silks are inferior to *Bombyx mori* in quality of thread and cloth; their only benefit is cheapness.

Antherea dernyi
Chinese oak silkworm from which the pale buff-colored Shantung silk is made. Native of Mongolia but now spread through Asia.

Antherea yamani
Japanese oak silkworm, imported to Europe in the 19th century to found the silk production industries of Britain and France. The eggs hatched before the oak buds burst and so culture failed.

Samia cynthia
First cultured species was developed in Japan but traditionally raised in Assam and Bengal. Several races cultured worldwide; used to produce silk in the fledgling industry in Brazil. The race Ricini is most important giving white or brick-red Eri silk.

Bombyx mori
The industrial source of silk, accounting for 85 percent of world production. It is superior to all others in all aspects of quality. The undyed silk is white or yellow.

SILK has been produced by man for thousands of years either by collecting cocoons of wild silk moths or by artificial rearing. The silk industry is thought to have started in China where the Empress promoted rearing of silk worms and reeling of silk threads 4,500 years ago. Probably small-scale production also occurred in other Asian countries. The Chinese guarded the secrets of silk culture but they spread gradually from China to Japan.

The most productive species (*Bombyx mori*) no longer exists in the wild. The larvae and adults are white; all color has been lost over the thousands of years of intensive artificial rearing. The adults have also lost the ability to fly.

Normally this moth has one generation per year and so farming was a seasonal occupation in peasant agriculture. The eggs were overwintered and allowed to hatch at any time from the beginning of April until the end of May, depending on when, in the producing country, the buds of the mulberry, its natural host plant, burst. Eggs were kept cool until foodstuff was available. Then they were incubated at a higher temperature, often in a bag around the rearer's neck, to stimulate hatching.

Silk moth rearing on a large commercial scale, principally by the Chinese and Japanese, was made possible by breeding various strains of "multivoltine" silk moths—strains having more than one generation per year. Sericulture, the rearing of silk moths, then became a full-time industry and became more productive as hybrid crosses between different strains gave much more silk.

Continual cultivation of the insects requires mulberry leaves all year. Mulberry plants from different countries and continents are used as well as diets which contain an extract of mulberry to stimulate the larvae to feed on them. Large-scale silk worm production requires large-scale mulberry production and one of the major requirements of the industry is quick maturing and very productive trees. Various cultivated mulberry strains have been developed for this, especially bush and shrub types. These give earlier maturity and allow the leaves to be picked more efficiently.

In all rearing systems cleanliness is essential. Many times a whole season's production has been wiped out by disease. Disease brought silk production to an end in France in 1956, although a high-quality industry has been revived, partly as a tourist attraction, in the Basses-Cévennes.

Several thousand eggs are spread on each rearing tray in a fumigated chamber and covered with fragmented mulberry leaves. Newly hatched larvae crawl onto the leaves to eat them. Each larva consumes some 1 oz

World Importance of Silk Moth Harvesting

Silk production occurs worldwide since silk moth farming is artificially controlled, but as a labor intensive industry it is suited best to an underdeveloped economy. That is why a large proportion of world output has shifted from Japan to China and various developing countries over the last 50 years. About 65 percent of world production is now by developing countries, and an industry only survives in Japan because of measures taken by the Japanese government.

Attempts to establish silk industries in Europe have been made several times. These were most successful in the United Kingdom and France in the late 19th century but failed because of the seasonality of the silk moths and difficulties in rearing foodstuffs. Production still occurs in Europe but is a minimal proportion of the world output.

World output has remained fairly constant throughout the 20th century. The total estimated world production from *Bombyx mori* moths in 1938 was 56,000 tons, in 1978 50,000 tons and in 1980 54,000 tons. There is also considerable production from moths of other species in India and Sri Lanka. China produced 12,000 tons of non-*Bombyx* silk in 1982. A considerable proportion of world output is used by the domestic economies of producing nations. In 1984 raw silk from *Bombyx mori* cost $13.50 per lb ($30 per kg) and processed silk such as a dyed six-thread yarn cost $32 per lb ($70 per kg). So the value of silk thread produced in 1984 by the world's silk moth industries was approximately $36 million.

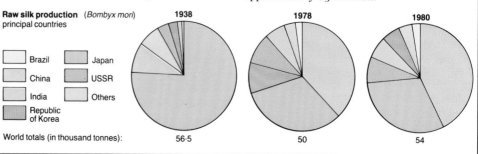

Raw silk production (*Bombyx mori*) principal countries

☐ Brazil
☐ China
☐ India
☐ Republic of Korea
☐ Japan
☐ USSR
☐ Others

World totals (in thousand tonnes):

1938	1978	1980
56·5	50	54

▲ ▼ **Controlled life cycle.** ABOVE A few silk moths are allowed to reach the adult stage and destroy their valuable cocoons by emerging to breed. Eggs laid by adult females will hatch out as larvae BELOW to feed on mulberry leaves for 4–6 weeks before spinning their own cocoons. Most of the pupae within are then killed by boiling or baking before the silk is unwound and run onto bobbins. These are *Bombyx mori*.

(30g) of leaves supplied in small amounts up to six times per day over 4–6 weeks. Following each molt the trays are cleaned and a synchronized culture is obtained, as late-molting larvae are removed with the cast skins and feces.

When mature, the larvae cease feeding and wander from the foodstuff into straw or similar material, provided after the last larval molt, in which they spin their cocoons. Cocoons are left for a week to harden before harvesting. A few are retained to mature. From these, adults emerge after about one month. They mate and the females lay eggs. These eggs are either stimulated to hatch or kept to over-winter depending on the type of culture.

After harvesting, the living pupae within the harvested cocoons are killed either by boiling or baking. This prevents the adults emerging and dissolving the continuous silk thread; it also makes the pupa shrink away

from the silk thread. The cocoons can be stored in this condition until the next stage of processing or they can be marketed.

To obtain the silk thread the cocoons are boiled and mechanically agitated to remove loose bits of silk and other debris. This process also loosens the individual continuous thread of each cocoon. Thread from several cocoons is then collected together and the silk fibers unwound—all by machine nowadays. This produces hanks of silk which are run onto bobbins. The silk on three of these bobbins is then wound together to form a silk thread. Finally the silk threads are mechanically tensioned by winding them from 150–1,000 times to produce a thread whose thickness, weight, fineness and mechanical properties will depend on its final use. The silk may be dyed at this stage, and the processed thread can be woven into any of a multitude of clothing and commercial products.   EHi

HUNTERS
AND GATHERERS

▶ **The fisherman's feathered friend.** ABOVE
Setting out for a night's fishing with
cormorants on the river Li-karst in Yang-shuo,
China. The birds are fitted with collars to
prevent their swallowing the fish. When
released they dive into the water, returning
only when they have caught a fish. BELOW
Cormorant disgorging catch into a Chinese
fisherman's basket.

DOGS were the first animals to become
associated with man through hunting.
Wild dogs first competed with man at kills
but later discovered that if they cooperated
with man they could share in the kill. A pattern of coordination between man and dog
evolved. The relationship was easily extended to the specialized training of dogs for
hunting, retrieving and herding.

Other animals have also been trained to
hunt for man. The **cheetah**, would probably
have been a domesticated animal today had
it not been for its reluctance to breed in captivity. The ancient Egyptians and Assyrians
first hunted with tame cheetahs, but it was
in India in the 15th–18th centuries that the
sport of coursing cheetahs reached its zenith
of popularity. The cheetah was held by a collar and chain and blindfolded until a black-
buck was sighted. It was then released and
a short chase ensued. If successful the
cheetah was rewarded with blood from the
kill. Cheetahs were also used to hunt
gazelles by the Arabs and Abyssinians and
coursing was common in Africa by the end
of the 16th century. The **caracal** was also
coursed.

Unlike the cheetah, the **ferret** breeds easily
in captivity and has been domesticated for
at least 2,000 years. It seems likely to be
descended from either the European polecat
(*Mustela putorius putorius*), or the steppe
polecat (*M. eversmani*). There are now two
color forms, the albino and the brown,
which is a wild-type polecat-ferret. Pliny in
the 1st century described the use of ferrets
to combat a plague of famine-causing rabbits in Spain. The Romans introduced rabbits to Britain from Spain and then brought
the ferret to work the warrens.

Ferrets have been an indispensable part
of the European poacher's equipment for
centuries and today they are still used to
control rabbit populations in the traditional
manner. A well-frequented warren is found
and all the exits identified. Small nets are
then placed over all the holes and a ferret
is introduced. Rabbits bolting from the burrows become entangled in the nets and are
easily killed. The holes may be left
uncovered and the fleeing rabbits shot.

Sometimes the ferret will kill below
ground, fall asleep and not return to the surface. A large, aggressive line ferret is then
sent down on the end of a rope to chase the
other out. Contrary to popular opinion,
well-treated ferrets become very tame and
require little training in order to work with
great enthusiasm.

In a similar manner, the **grison**, a wild
South American relative of the ferret, is used
by Peruvian hunters to flush the Plains
viscacha from its burrows. Another
mustelid, the **Yellow-bellied weasel**, is kept in
houses in Nepal as a mouser, as are many
species of **mongoose**.

The fisherman has been aided in his work
by a variety of animals over the centuries
including otters, cormorants, ospreys and
dolphins. The use of **otters** for fishing, generally the Oriental short-clawed otter or river
otters, originated in South China during the
Tang dynasty (606–916AD). Men living
near the river each had their own otter
which lived in a den in the bank. When
released the otter would catch large numbers of fish, most of which were taken by
the man. Later, otters were trained to fish
and retrieve on command. They were used
in this way in India during the 19th century
where fishermen kept them tethered to their
boats and put leather straps over their
canine teeth to prevent them from spoiling

the fish. More sophisticated methods of using otters were also devised, making the well-trained otter a valuable asset. On the upper Yangtze River, otters were trained to work with fishermen using nets. The otter was muzzled and then dropped beside a large dip net into which it drove the fish. In Europe, the use of fishing otters became widespread during the 16th and 17th centuries and the practice is referred to in a document from 1618 in England, concerning the payment of £66 13s 6d to the "Keeper of his Majesty's cormorants, ospreys and otters." Various methods of training otters have been suggested, but as they become bored easily and can inflict nasty bites, it requires much skill and patience. Some pet otters have taken to retrieving game both on land and water and there is one report of a tame otter in Hampshire, England, which lived with a pack of otterhounds and hunted and killed its own kind.

As with the otter, the practice of using **cormorants** for fishing originated in the east and later became popular in the west. They were first domesticated in Japan in the 5th century and spread to China. It was the Jesuits who brought the practice to Europe where it became established in the 16th and 17th centuries. Today two species are still used for fishing in Japan: the Common or Great cormorant and the Japanese or Temminck's cormorant.

It is not only carnivorous animals which have proved useful to man. The **pig**, with its high intelligence and keen sense of smell is closer to the dog in character than any other domestic animal and has served man in several dog-like ways. It has long been used to find and dig up truffles in the Perigord region of France. The truffle is a large fungus which grows at depths of 7.5–15in (30–60cm) under the soil. Normally a sow is used being easier to handle than the boar and

Docile Dolphins

The mutual fascination between man and dolphin has existed for thousands of years and has often developed into cooperative fishing. Pliny the Elder in AD70 observed fishermen in France calling dolphins to drive a shoal of mullet into shallows. There, trapped between men and dolphins, large numbers could be caught.

The Ganges dolphin, which inhabits the Ganges, Brahmaputra and Indus rivers, has long been trained to herd fish shoals toward the nets of fishermen. It becomes so tame that it often swims between the legs of bathing pilgrims.

In Australia, aborigines have cooperated with wild dolphins for centuries. When a shoal of mullet is sighted, they beat the surface of the water with their spears. This brings the dolphins inshore to drive the mullet into the shallows, where both men and dolphins can share in the catch.

The Amazon dolphin, a river dolphin found in the Upper Amazon, is highly regarded by the local Indians, as in its presence bathers are protected from the rapacious piranha fish. The Indians have developed close fishing partnerships with individual animals. The man calls his dolphin by whistling and it then feeds across the river from his canoe. The dolphin's activities drive the fish towards the man and the man's drive the fish towards the dolphin, resulting in more successful fishing for both. It is an indication of the dolphin's intelligence that it does not need to be captured and trained in order to assist man, but will voluntarily enter into a long-term relationship which is profitable to both parties. However, due to the increased sophistication and scale of modern fishing techniques the dolphin is now more often regarded by fishermen as a competitor than an ally.

more readily attracted to them, if it is true, as suggested, that the truffles produce sexually stimulating chemicals which resemble pheromones produced by the boar. She is muzzled to prevent her eating the fungus and may be accompanied by her piglets who learn to truffle from her. Man simply exploits the pig's natural rooting behavior and taste for truffles to acquire this expensive delicacy.

In the 11th century William the Conqueror made the New Forest in England a royal game reserve and prohibited dogs above a certain height to prevent their use for hunting. The Saxon peasantry retaliated by training pigs to flush out and retrieve game. These "hunting pigs" continued in use up to the 15th century. They were found to be a match for the finest hunting dogs in their ability to point, flush and retrieve game and there are even reports of pigs being used to round up cattle.

The **honey-guide** is an unusual example of an animal which assists man in finding food on a strictly voluntary basis. It is a small brown bird found in Africa and related to the woodpeckers. Two species are known to guide both man and the ratel, or Honey badger, to bees' nests. The bird calls insistently and flutters its wings to attract the predator's attention to the location of an active nest. The predator then opens it up to obtain the honey, at the same time exposing grubs and wax for the bird to eat.

Unlike the honey-guide the Pig-tailed **macaque** doesn't offer its services; it has to be taken captive and trained to work for man. The Pig-tailed is the largest of the macaque family and is widely distributed from Burma to Sumatra. It is commonly taken when young, tamed and trained to climb trees and throw down ripe fruit and coconuts. Mature males become aggressive and can be dangerous; so only young animals and females are used. The practice has proved so successful and is becoming so widespread that a "school" has been established in Sumatra.

Other primates have also made themselves useful to man in a variety of ways, although not on any large scale. There is a remarkable report of a Chacma **baboon** which worked as a goat herd in South West

► **Hunting grouse in the Scottish Highlands.** This one-and-a-half-year-old female Peregrine falcon has been trained to return to the falconer when the lure in his ungloved hand is twirled on a cord. Still most widely practiced in the Asian countries where it originated, falconry was one of the foremost sports of European aristocrats between the 9th and 16th centuries.

► **Catching fish in Bangladesh** BELOW. Otters were first used as fishermen's helpers in China. Sometimes the man simply takes some of the fish which his animal catches naturally, but otters have been trained to fish on command, even as muzzled herders driving fish into nets. The canine teeth may be covered with leather straps to avoid damage to the catch. Use of European species for fishing was widespread in the 16th and 17th centuries.

▼ **Rooting for truffles in France.** The Perigord region of France is famous for its truffles, a prized underground fungus, often growing among the roots of oak trees and impossible to find without animals trained to the scent. In the Perigord, pigs are used. Up until 1930, dogs were regularly used in Wiltshire, England, where the truffles still grow, mostly uncollected, among the roots of beech trees.

Africa. The baboon, a young female, was left in a corral with the goats for a few weeks until she regarded them as her troop. Soon she accompanied them when they were allowed out to graze, watching over them all day and bringing them back to the corral in the evening. She could recognize individual goats and would carry straying kids back to their own mothers, even placing them on the udder. Other baboons were also trained and worked in a similar way.

Falconry is the sport of using **falcons**, **hawks** and sometimes **eagles** to capture and kill wild game. It originated in the East as early as 1200BC and today the art is still practiced most extensively in India, Pakistan and Saudi Arabia. Falconry was introduced to Europe in the 9th century and became the foremost sport of the aristocracy until its decline in the 16th century. Today it is practiced on a small scale in Europe and the United States, where falcons are sometimes used to clear potentially hazardous flocks of birds from airfields.

Two groups of birds are used, the long-winged true falcons (eg Peregrines, merlins and sakers) and the short-winged hawks (eg goshawks and sparrowhawks). The falcons are used for hunting game such as grouse and partridge in open country, the falcon being released before the quarry is put up (forced out of cover) in order that it can "stoop" down to kill. In contrast the hawks are flown in woodland and scrub to hunt rabbits or small birds and they are released after the quarry has been put up. They use a horizontal rather than vertical dash to make the kill. Training of hawks and falcons can be done with both young and old birds taken from the wild. Initial taming involves feeding the bird on the gloved hand until it accepts the falconer. Later it is trained to feed from the lure, a padded weight on a cord to which a few feathers are attached, and then to fly to the lure when it is whirled on the cord. Eventually it can be taken hunting and the lure used to make the bird return to the falconer.   MAB

FARMING ENDANGERED SPECIES

MANY species are threatened with extinction because of commercial exploitation in the wild. One way of relieving this pressure is to establish domestic populations of the same species and supply commercial demand from these instead. Such farming may also promote further conservation measures for the wild populations—for example, prevention of habitat destruction—through demonstrating the economic value of a continued existence for the species.

Farmers of endangered species should be controlled by legislation to protect the wild populations from excessive exploitation, but it is very much up to the discretion of individual governments what sort of farming to permit. On an international level, commerce in endangered species is strictly regulated by the Convention on the International Trade in Endangered Species of Wild Fauna and Flora (CITES). In all, 87 countries were party to this Convention in 1984.

CITES is expressly designed to control international trade which may be contributing to a species' decline, for example, trade in rhino horns from Africa and Asia which are sold in the Far East for medicinal purposes. CITES divides threatened species into three categories: Appendices I, II, and III to the Convention. Appendix I lists those species which are most severely threatened with extinction. International commercial trade in Appendix I animals or their products is not permitted, except under exceptional circumstances.

CITES does recognize that there are some populations of Appendix I species that can tolerate limited trade and that there are some populations that can support exploitation even though elsewhere the species is highly endangered. One example is the Salt-water crocodile. The species is listed in Appendix I, except for the population in Papua New Guinea, where it is numerous. This population is listed specially in Appendix II which permits limited international trade.

Exemptions from the trade embargo in Appendix I are also possible for captive-bred animals. Under CITES regulations if an animal is the result of sexual reproduction occurring in a controlled environment, it may enter into international trade like an Appendix II species. The animal must not only be born in captivity but conceived in captivity, otherwise it is not eligible for the exemption.

The purpose of this tight definition of captive breeding is to encourage independence from the wild. To be considered a farm, an operation must maintain and breed individuals whose offspring actually provide the traded commodity. Many so-called farms are not actually independent and still take eggs or animals from the wild. Turtle farms often rely on collecting eggs laid by wild turtles which are then hatched and reared in captivity before being slaughtered for their meat and shells. This means that the wild turtles are exploited at two stages—as eggs and as adults which continue to be taken directly from the sea. This sort of operation is more properly called a ranch.

An example of a true farm is provided by the raising of Dwarf musk deer in China. Musk is the secretion of an abdominal gland in the male and it is highly sought after by the perfume and medical industries. Males are killed to get the musk, and musk deer species of six countries are listed on Appendix I, but the Chinese appear to have developed a technique of extracting the musk without killing the animal. The male

▲ Domestication of endangered species. Successful breeding and farming of the diminutive Dwarf musk deer (*Moschus berezovskii*) has been achieved in China, where males are "milked" up to 14 times for their musk, used in oriental medicines and western perfumes. Wild musk deer in Tibet are threatened with extinction by hunters who kill for a single extraction of this product. Shown here is a female Dwarf musk deer, lacking the distinctive tusks of the male.

◀ Destined to be skinned, a Saltwater crocodile feeds at a crocodile ranch in Darwin, Australia. Similar ranches have been established in several countries, causing concern for the conservation of this threatened species (*Crocodylus porosus*), because young are taken from the wild. In Papua New Guinea, where the Saltwater crocodile and the New Guinea crocodile (*C. novaeguineae*) are relatively numerous, crocodile ranching and conservation are acknowledged to be successfully integrated by a national program earning the country US$2 million annually from about 50,000 skins. Small crocodiles are caught in the wild and kept in family farms or collective village holdings before being sold to private or government ranches which eventually sell skins to a government agency for export. Villagers who once considered crocodiles as pests or threats to be eliminated are now concerned to promote their future existence, to guarantee continued cropping from the wild.

then regenerates the musk and it can be harvested again, up to 14 times.

Under these sorts of conditions it becomes feasible as well as convenient to hold the species in captivity—in commercial terms, to set up a farm. China has recognized the potential and other countries such as Nepal are following suit. Although Dwarf musk deer are on Appendix II, in theory the techniques developed could be extended to Appendix I species of musk deer and commercially viable operations could be established. Captive stocks could supply a large part of the musk in trade which would take the pressure off the wild, presuming that the trade was not stimulated by the continued or increased availability of the product.

Given these desirable consequences it might seem surprising that there are so few farms for endangered species in existence, but the obstacles to the establishment of farms for wild animals are considerable. The cultivation of any non-domestic species, particularly an endangered one for commercial purposes requires initiative, skill and considerable financial backing. The economic realities of farming compounded by a lack of the expertise and technology necessary to breed exotic animals on a commercial basis are major problems and more than anything else may be the reason for the relative scarcity of true farms.

Farming is a risky business even with traditional stocks—pigs or poultry, for example. It is capital- and often labor-intensive, and while poultry have a relatively assured and local market—there will always be a demand for chicken—crocodile farmers, for instance, have no such security.

Endangered species have an intrinsic value born of scarcity. This often means that their products are regarded as luxury commodities—reptile skins are used for relatively expensive leather goods. The market is subject to radical fluctuations in price and demand. The crocodile farmer may suddenly find that crocodile wallets are no longer in fashion and there are not enough buyers for his skins. The markets for such luxury goods are usually in North America, Europe, and the Far East (primarily Japan, Hong Kong and Taiwan). The problems and expense of shipping products this distance are considerable and add to the financial risks.

Further obstacles are presented by opponents of commercial exploitation. While farms for threatened species may reduce the pressures on wild populations, equally the continued availability of products may support, even increase, the demand.

Continued availability may also make it difficult to regulate trade. Some supplies will come from approved sources. Others may be illegal, yet it may not be possible to identify which come from where.

Few bona fide farms exist. Ranching operations represent the majority of commercial enterprises involving endangered species. They have fewer intrinsic financial problems than farms because they are not forced to maintain independent breeding stocks. But ranching encounters difficulties in justifying its continued exploitation of wild populations of Appendix I species.

Conditions in a ranching system are extremely flexible, making it very difficult to construct a hard and fast definition which protects the wild populations in the way that the captive-bred clause of CITES does. The growth of ranching operations as a form of commercial exploitation of endangered species is thus likely to be a major issue in international conservation.

Both farming and ranching have advantages to offer to conservation but both must be carefully regulated to protect wild populations from excessive economic zeal which merely precipitates the decline of species. The potential profits to be gained are not purely monetary and it is important to remember that the continued existence of a species is in itself a valuable asset.　AD

BIBLIOGRAPHY

The following list indicates key reference works used in preparation of this volume and those recommended for further reading.

Banks, S. (1979) *The Complete Handbook of Poultry Keeping*, Ward Lock, London.

Bath, D.L., Dickinson, F.N., Tucker, H.A. and Appleman, R.D. (1978) *Dairy Cattle* (2nd edn), Lea and Febiger, Philadelphia.

Baxter, S. (1984) *Intensive Pig Production*, Granada, London.

Brown, E.E. and Gralzek, J.B. (1980) *Fish Farming Handbook*, Avi, Westport, Virginia.

Bundy, C.E., Diggins, R.V. and Christensen, V.W. (1984) *Swine Production* (5th edn), Prentice Hall, Englewood Cliffs, New Jersey.

Clutton-Brock, J. (1981) *Domesticated Animals from Early Times*, Heinemann, London.

Cockrill, W.R. (ed) (1974) *The Husbandry and Health of the Domestic Buffalo*, FAO, Rome.

Crabbe, D. and Lawson, S. (1981) *The World Food Book*, Kogan Page, London.

Delacour, J. (1977) *The Pheasants of the World*, Spur Publications, Hindhead, Surrey.

Devendra, C. and McLeroy, G.B. (1982) *Goats and Sheep Production in the Tropics*, Longman, London.

Eltringham, S.K. (1982) *Elephants*, Blandford, Poole, Dorset.

Epstein, H. (1969) *Domestic Animals of China*, Commonwealth Agricultural Bureaux, Farnham Royal, England.

FAO (1984) *FAO Production Yearbook 1983*, vol. 37, FAO, Rome.

Friend, J.B. (1978) *Cattle of the World*, Blandford, Poole, Dorset, England.

Gaskin, D.E. (1982) *The Ecology of Whales and Dolphins*, Heinemann, London.

Gautier-Pilters, H. and Dagg, A.T. (1981) *The Camel – its Evolution, Ecology, Behavior and Relationship to Man*, University of Chicago Press, Chicago.

Geist, V. and Walther, F. (eds) (1974) *The Behavior of Ungulates in Relation to Management*, IUCN, Morges, Switzerland.

Hafez, E.S.E. (1975) *The Behavior of Domestic Animals* (3rd edn), Baillière Tindall, London.

Kilgour, R. and Dalton, C. (1984) *Livestock Behaviour*, Granada, London.

Mace, H. (1976) *The Complete Handbook of Bee Keeping*, Ward Lock, London.

Mason, I.L. (1969) *A World Dictionary of Livestock Breeds, Types and Varieties* (2nd edn), Commonwealth Agricultural Bureaux, Farnham Royal, England.

Owen, J.B. (1976) *Sheep Production*, Baillière Tindall, London.

Pitcher, T.J. and Hart, P.J.B. (1982) *Fisheries Ecology*, Croom Helm, London.

Politiek, R.D. and Bakker, J.J. (1982) *Livestock Production in Europe*, Elsevier, Amsterdam.

Ponting, K. (1980) *Sheep of the World*, Blandford, Poole, Dorset, England.

Reay, P.J. (1979) *Aquaculture*, Arnold, London.

Reed, C.A. (ed) (1977) *Origins of Agriculture*, Mouton, The Hague.

Rouse, J.E. (1969) *World Cattle*, 3 vols, University of Oklahoma Press, Norman.

Spedding, C.R.W. (1975) *The Biology of Agriculture Systems*, Academic Press, London.

Spedding, C.R.W., Walsingham, J.M. and Hoxey, A.M. (1981) *Biological Efficiency in Agriculture*, Academic Press, London.

Stansfield, M. (1983) *The New Herdsman's Book*, Farming Press, Ipswich, England.

Tazima, Y. (1964) *The Genetics of the Silkworm*, Logos Press, London.

Warner, G.F. (1977) *The Biology of Crabs*, Elek, London.

Yerex, D. (1979) *Deer Farming in New Zealand*, Agriculture Promotion Associates, Wellington, N.Z.

Zeuner, F.E. (1963) *A History of Domesticated Animals*, Hutchinson, London, Harper and Row, New York.

Zhigunov, P.S. (1968) *Reindeer Husbandry* (translated from the Russian), Israel Program for Scientific Translators, Jerusalem.

GLOSSARY

Aboriginal population a breeding group of animals which retains the same form as ancestors which constituted the original native population of that location, as opposed to DOMESTIC ANIMALS and populations outbred to strains introduced by man.

Action the way a dog or horse moves. In some breeds, this forms part of the breed standard and so is judged at shows.

Adaptation MUTATION that gives an organism a survival advantage in its ENVIRONMENT. NATURAL SELECTION favors the survival of individuals whose adaptations adjust them better to their surroundings than the other individuals.

Adaptive radiation The pattern in which different forms of species develop from a common ancestor by evolving differences suiting them to occupy different ecological NICHES.

Adrenal gland an organ of HORMONE secretion in vertebrates. There is a single pair, one near each kidney, in man and other mammals. Each gland has two components, the medulla and cortex, distinct in function but closely fused together. The medulla secretes adrenaline and noradrenaline and its activity is controlled by the sympathetic nervous system. The cortex secretes various sterioid hormones. Cortical hormone (especially glucocorticoid) secretion is controlled by a pituitary hormone.

Adult a full developed and mature individual capable of breeding.

Agouti a grizzled coloration resulting from dark hairs pale at the tip, or from alternate light and dark barring of each hair. The basic coat color of many species such as cats, rabbits, mice and rats.

Albinism absence of pigmentation in the skin and hairs and eyes, so the animal is white, usually with red eyes. Due to recessive mutation.

Albino an individual animal which displays ALBINISM.

Alfalfa (*Medicago sativa*) leguminous crop, widely grown for HAY and FORAGE: lucerne.

Allergy sensitivity to foreign substances.

Amble a GAIT used by certain breeds of horse where both feet on the same side of the body are lifted together.

Amino acids organic compounds containing both basic amino (NH$_2$) and acidic carboxyl (COOH) groups. Amino acids are fundamental consituents of living mater—some hundreds of thousands of amino acid molecules are combined to make each PROTEIN molecule.

Analgesic pain-killing drug.

Ancestor an individual from which an animal is descended. In EVOLUTION, it refers to a living or fossil species from which a present-day species is believed to have descended. Most domestic animals are derived from wild ancestors whose wild descendants are considered to be of the same species as the domesticated descendants.

Aquaculture culture of animals or plants associated with water.

Arid habitat an area of land with a low annual rainfall and covered by sparse vegetation adapted to such environmental conditions.

Artificial insemination artificial injection of semen into female.

Artificial selection the process whereby animals allowed to produce offspring are chosen by humans, in contrast to NATURAL SELECTION where biological fitness determines the success of reproduction.

Ass donkey.

Aviculture the rearing and breeding of birds.

Awn hairs medium to long hairs of mammals. They are thinner than the GUARD HAIRS and have a characterisitc thickening in diameter near to the tip which then tapers to a fine point.

Backbreeding the mating of an individual with one of its parental types.

Bacon cured back and sides of pig.

Bantam small breed or variety of domestic fowl. Often a miniature of a standard breed.

Barrel the trunk of a four-legged animal.

Battery cage a small cage, usually of wire mesh, designed to hold hens in captivity for egg production.

Beef flesh of ox, bull or cow.

Binocular vision a form of vision typical of birds and mammals in which the same object is viewed simultaneously by both eyes: the coordination of the two images in the brain permits precise perception of distance.

Bloodline the line of descent from ancestors. Often used to describe individuals which share a common ancestry and are specially bred to retain desired characteristics.

Boar uncastrated male pig.

Bonemeal ground-up bone used as fertilizer.

Bovine pertaining to cattle. Also (noun) an individual cow, bull, steer etc; also individual of any cattle species, including buffalos, bisons, yaks, mithuns, etc.

Brassica plants belonging to the genus *Brassica*, eg cabbage, rape, mustard.

Breed to produce offspring. Also, to cross selected individuals to produce offspring with desired characteristics. Also (noun) a race or strain whose members, when crossed, produce offspring with the same characteristics as the parents.

Breeding the rearing and crossing of animals or plants so as to change the characteristics of future generations.

Breeding stock animals that are used for BREEDING.

Brindled having inconspicuous dark streaks or flecks on an otherwise gray or tawny background.

Brisket the part of the animal's lower chest lying between the forelegs.

Brood to incubate or sit on eggs or young.

Broodstock birds to use for BREEDING.

Broody of hen wishing to sit or INCUBATE.

Browse to feed on shoots and leaves (as distinct from GRAZE).

Buck the male of species such as rabbit, hare, deer, antelope.

Bull uncastrated male of any bovine animal.

Bullock a young bull. Also CASTRATED bull, especially young castrated bull: STEER.

Burro donkey.

Buttermilk the acidulous milk which remains after butter has been churned out.

Byproduct thing produced incidentally in manufacturing something else.

Calf BOVINE young, especially young of domestic cow.

Camouflage the disguise of an animal's appearance in its natural HABITAT through the coloration and patterning of its coat, skin or feathers.

Capon castrated cock.

Carbohydrate organic compound of general formula Cx(H$_2$O)y: eg sugars, starches, cellulose. Carbohydrates play an essential part in the metabolism of all organisms.

Carcass the dead body of an animal.

Carnassial teeth opposing pair of teeth especially adapted to shear with a cutting (scissor-like) edge: in living mammals the arrangement is unique to carnivores and the teeth involved are the fourth upper premolar and first lower molar.

Carnivore any meat-eating organism. Alternatively, a member of the order Carnivora, many of whose members are carnivores.

Carrion dead flesh, often having undergone some putrefaction.

Cartilage hard, elastic tissue found in animals. In some fish, such as sharks, the whole skeleton is made of cartilage rather than bone.

Castrate remove testicles, geld.

Cereal a general term for grains produced by members of the grass family. eg wheat, rice, barley, millet, which are commonly consumed by people.

Cheese the curd of milk, coagulated by rennet, separated from the whey and pressed into a solid mass.

Cholesterol complex organic alcohol related to steroids, found in an animals studied, but not in plants or most bacteria. An important constituent of animal cell membranes.

Chromosomes thread-shaped body structures in the nuclei of animal and plant cells consisting of nuclei acid (DNA and RNA) and PROTEINS. They carry the genes which control inherited characteristics.

Class a major category in taxonomical classification ranking above ORDER and below PHYLUM.

Clavicle the collar bone.

Cloaca terminal part of the gut in some animals, eg birds, rabbits, into which reproductive and urinary ducts open and from which there is only one opening, the cloacal aperture.

Commensal animal or plant species living and feeding in the same area as another: in the strictest sense, one derives benefit from the other but neither suffers on account of the association.

Conformation the general body form of an animal.

Core body temperature the temperature of an animal's body most distant from the skin and extremities. It generally fluctuates little and is usually the highest temperature of the body.

Courtship period of activity when males and females seek to attract each other prior to mating.

Crepuscular tending to greater activity during twilight (at dawn and dusk).

Crossbreed the offspring of the mating of two different BREEDS.

Croup the rear part of a mammal's back, immediately in front of the base of the tail.

Dam a female parent.

Decomposition the action of breaking up: decaying, rotting.

Dental formula a convention for summarizing the dental arrangement. The numbers of each type of tooth in each half of the upper (above the line) and lower (below the line) jaw are given in the order: incisor (I), canine (C), premolar (P), molar (M). A typical example for Carnivora would be I 3/3. Cl/1 P4/4, M3/3 = 44. The final figure is the total number of teeth.

Dentition the arrangements of teeth characteristic of a particular species.

Desi Indian cattle and buffalos of no breed.

Dewlap loose, pendulous skin under the throat of some cattle and dogs.

Dilute a paler version of a basic coat or feather color.

Display any conspicuous pattern of behavior that can convey specific information to others. Usually to members of the same species; can involve visual and/or vocal elements, as in threat courtship or "greeting" displays.

Distribution the geographical areas where an animal is found.

Divergence the separation over time of a SPECIES into breeding POPULATIONS of different appearances. These may become RACES, SUBSPECIES or ultimately SPECIES in their own right.

Docility tractability, tameness, or submissiveness of an animal.

Doe a female rabbit, hare, deer etc.

Dog a male member of the family Canidae (including the domestic dog); also a male ferret.

Domestic animal an animal living in proximity to man which has undergone the process of DOMESTICATION.

Domestication the process of altering the behavior and physiology of initially wild populations through ARTIFICIAL SELECTION for such characteristics as docility, high yield of desired products or ease of breeding.

Dominant animal an individual which as part of a social group has a higher status than other members. While dominance may originally be established by aggressive encounters, these are often not involved in its maintainance. Dominant animals have first access to resources such as food, space and mates.

Dominant gene, dominant trait a gene and its accompanying trait which manifest themselves in a similar fashion regardless of the presence or absence of the corresponding RECESSIVE gene at a matching site of the paired CHROMOSOME.

Double muscling genetically produced increase in cattle in the number of muscle blocks and hence the size of muscles.

Down feathers the first feathering of young birds. Also, the fine soft under-plumage of birds.

Down hair the short, soft densely packed type of hair lying closest to the skin.

Draft animal one bred for strength and used to draw carriages, wagons or farm implements.

Drake male duck.

Drive the time when shoals of fish form for spawning. Males chase ripe females which will shed their eggs in a suitable place. The males then release MILT, which contains sperm, to fertilize the eggs.

Drover a person or animal that helps drive a herd or flock of livestock.

Dub cut off the comb and wattles of a cock.

Ecology the study of interrelationships of plants and animals in their natural environmental setting. Each species may be said to occupy a distinctive ecological NICHE.

Ecosystem a unit of the environment within which living and nonliving elements interact, eg a lake, inhabited by a community of animals and plants.

Embryo the developing offspring of an animal before birth or hatching from an egg.

Endemic species a species of plant or animal unique to a particular region, country, island, etc.

Entire male uncastrated male.

Enviroment collective term for the conditions in which an organism lives influenced by temperature, light, water, other organisms, etc.

Enzyme a PROTEIN which is a catalyst of biochemical reactions. There are many different kinds each kind directly promoting only one or a very limited range of reactions.

Estrus the period in the reproductive cycle of female mammals at which they are attractive to males and receptive to mating. The period coincides with the maturation of eggs and OVULATION.

Eviscerate to take out the entrails of: to disembowel, to gut.

Evolution the process by which species have developed to their present appearance and behavior through the action of NATURAL SELECTION in determining the survival of those individuals most suited to their ENVIRONMENT.

Ewe a female sheep.

Ewe neck a concave curvature of the topline of the neck of a dog or horse.

Expression the degree to which a GENE'S effects are evident in an animal. Factors such as ENVIRONMENTand the presence of other genes can mask or modify a gene's expression or expressivity in terms of final appearance.

Extensive farming system of farming in which animals are kept in fields or open country rather than in buildings or yards with little space per animal.

Family a taxonomic division subordinate to ORDER and superior to GENUS.

Fat substance which can be extracted from tissues by organic solvents such as ether but not water, and stored in adipose tissue. True fat is a compound of glycerole and fatty acids. Fats form a potential energy source.

Fatty acids organic aliphatic acid. Biological ones usually have long straight chains and an even number of carbon atoms.

Fawn the young of Fallow deer and certain other deer species.

Fecundity potential for producing large numbers of offspring. Fertility.

Feral living in the wild (especially the descendants of domesticated animals).

Fertility capability to produce offspring.

Fiber support tissue in plants which may form roughage in animal diet. Also strand of wool or other mammalian skin covering.

Fleece the woolly covering of a sheep or similar animal.

Flense to strip blubber from a whale or seal.

Flock book records of individual history and breeding of captive fowl.

Fodder feed for domestic HERBIVORES.

Follicle a small sac, eg, ovarian follicle, a mass of ovarian cells that produces an ovum; hair follicle, an indentation in the skin from which hair or feathers grow.

Forage food for horses, cattle etc, especially when acquired through BROWSING OR GRAZING.

Foundation breed one or two or more breeds contributing substantially to the establishment of a later breed.

Fowl domestic chicken. Also any of domestic duck, goose, turkey. Also any bird.

Free ranging in a domestic animal, having access to a more varied environment than a cage or pen. In a wild animals, a lack of territoriality and a wide individual range of movement or dispersal.

Game wild mammals or birds pursued, caught or killed by human hunters.

Gander male goose.

Gene the basic unit of heredity: a portion of a DNA molecule coding for a given trait and passed, through exact duplication at reproduction from generation to generation.

Gene pool the diversity of genetic information present within a group of breeding individuals. In small isolated breeding groups, the gene pool is usually small, but can be enlarged by OUTCROSSES to unrelated individuals.

Genetic purity degree to which individuals produce offspring true to their BREED.

Genetic resources the potential for genetic change in a species, breed or population.

Genetics the study of heredity and variation in animals and plants.

Genus (plural genera) a taxonomic division superior to SPECIES and subordinate to FAMILY.

Gestation the period of development of the mammalian fetus from conception to birth.

Gilt a young female pig up to the time when her first litter is weaned.

Glucose a six-carbon-atom sugar widely distributed in plants and animals; a unit in many CARBHOYDRATE compounds, eg sucrose, starch, cellulose, glycogen, and a major energy source in metabolism.

Gosling a young goose.

Graze to feed on growing grass.

Green algae primitive aquatic plants colored green by the presence of chlorophyll. There are both fresh and salt water species.

Guard hairs the long thick hairs of the coat of mammals. They protect the DOWN HAIRS, and often provide much of the color and pattern of the coat.

Habitat the ENVIRONMENT to which an animal is adapted.

Hair cornified elastic thread growing from an epidermal cell of the skin, characteristic of mammals.

Hardiness physical health and robustness, ability to withstand adverse conditions.

Harem group a social group consisting of a single adult male, at least two adult females and immature animals: a common pattern of social organization among mammals.

Haunch the flesh of the body on the side of the spine between the last ribs and the thigh.

Hay grass that has been cut and dried as feed for livestock.

Heat period during the reproductive cycle of female mammals whey they are sexually RECEPTIVE.

Heifer a young cow up to the stage at which her first calf is weaned.

Hemoglobin an iron-containing PROTEIN in the red corpuscles which plays a crucial role in oxygen exchange between blood and tissues in mammals.

Hen the female of birds or fish, especially of domestic fowl.

Herbivore an animal eating mainly plants or parts of plants.

Herb book records of individual history of BREEDING of cattle.

Hide the skin of an animal.

Hind female of Red deer and some other deer species.

Hindquarters see QUARTERS.

Hinny the offspring of a female donkey and a male horse.

Hob a male ferret.

Hock the joint in the hind leg of a quadruped between the true knee and its fetlock, the angle of which points backward. Also the knuckle end of a gammon of BACON.

Home range the area in which an animal normally lives, except for excursions and migrations.

Hormone organic substance produced in minute quantity in one part of an organism and transported to other parts where it exerts a specific effect, eg stimulating growth of a specific type of cell.

Humerus single long bone of the forelimb of mammals lying between the shoulder and knee.

Husbandry the practices of housing, managing, handling and rearing farmed animals.

Hybrid the offspring of parents which are not genetically identical.

Hybrid vigor a phenomenon frequently displayed by HYBRIDS where they may be healthier, fitter and/or larger than INBRED animals.

Hydrolyze to decompose by breaking chemical bonds of the type which occur in water.

Inbred produced by INBREEDING. Inbred animals are more likely to display genetic abnormalities.

Inbreeding breeding with close relatives, sometimes associated with reduced offspring survival.

Incubate to keep eggs warm so that development is possible.

Indigenous species a species that belongs naturally and historically to a given region.

Inguinal situated in the groin.

Inheritance genetic characteristics passed on from parents to offspring.

Inhibitor gene a GENE that masks or inhibits the action of another gene.

Interbreeding mating between genetically unrelated individuals. These may be different species or unrelated populations of the same species.

Introduction the spread of a species beyond its natural DISTRIBUTION, as a consequence of human action.

Iris colored tissue layer of the eye which surrounds the black pupil and is in turn surrounded by the white of the eye or the sclera.

Juvenile an individual no longer possessing the characteristics of an infant, but not yet fully ADULT.

Kemp a coarse or stout hair occuring among WOOL of an animal.

Lactation the secretion of milk from mammary glands. Also the period during which milk is being secreted.

Lactose sugar occurring in mammalian milk. Compound of a molecule of glucose and a molecule of galactose.

Lamb a young sheep.

Lambing percentage the ratio of surviving lambs to ewes in a flock, expressed as a percentage.

Lard the thick white fat of a pig, present in large quantity in some breeds.

Larva the juvenile form in which some animals, eg insects, amphibians, hatch from the egg.

Leather skin prepared for use by TANNING, drying or smoking.

Lie out in pregnant mammals, to separate from a social group preparatory to giving birth.

Line breeding the choice of breeding pairs over several generations so that a particular appearance or line is maintained. This line can be traced through PEDIGREE records.

Lineage the ancestry or PEDIGREE of an individual or breed.

Lipochrome a yellow organic pigment.

Litter a group of young animals brought forth at a single birth. Also, material used as bedding or for the dung of animals.

Livestock animals kept for economic reasons. STOCK.

Loin part of the body of a quadruped between the last rib and the hindlegs.

Longevity the length of life, lifespan.

Mahout an elephant-driver in India.

Mammary gland the milk-secreting organ of female mammals, probably evolved from sweat glands.

Mare a female horse.

Marine living in the sea.

Mask a colloquial term for the face of a mammal, especially of a dog, fox or cat. Also sometimes used for the face of a bird.

Maximum sustainable yield the greatest production possible within a given agricultural system, the greatest cropping possible of a naturally breeding population, such as a sea mammal or fish species, without depleting its numbers.

Medulla central part of an organ, eg adrenal gland, brain.

Melanin a black or dark brown pigment in the hair and the skin of many animals.

Mendelian genetics study of heredity and variation by breeding experiments as carried out by Mendel.

Metabolism the chemical processes occuring within an organism, including the production of PROTEINS from AMINO ACIDS, the exchange of gases in respiration, the liberation of energy from foods and innumerable other chemical reactions.

Midden a dunghill or site for the regular deposition of feces by mammals.

Milt mass of fish sperm.

Mineral any natural substance that is neither animal nor vegetable.

Mixed-use breed breed suitable for several products, eg cattle used for meat, milk and haulage.

Molt the replacement of old PLUMAGE in birds, or coat in mammals, by a new one.

Musk an odoriferous reddish-brown substance secreted in a gland or sac by male Musk deer.

Mutation a structural change in a GENE which may give rise to a new heritable characteristic if it occurs in one of the germ cells.

Mutton the flesh of sheep, as food.

Muzzle the part of the head of a mammal that projects in front of the eyes, including the nose, jaws and mouth.

Myxamatosis a viral disease of rabbits, usually lethal.

Natural selection the process whereby individuals with the most appropriate ADAPTATIONS are more successful than other individuals, and hence survive to produce more offspring.

Nectar the sweet fluid produced by flowers, especially as collected by bees.

Neoteny the retention of juvenile features in an adult animal.

Niche the role or status of a species within an ECOSYSTEM, defined in terms of all aspects of its lifestyle (eg food, competitors, predators and other resource requirements).

Nocturnal active at nighttime.

Nonfat solids components of milk which are neither liquid nor fat.

Nutrient a nutritious substance.

Omnivore an animal eating a varied diet including both animal and plant tissue.

Order a taxonomic division subordinate to CLASS and superior to FAMILY.

Outbreeding the mating of an individual with one that is unrelated, whether of the same or a different breed. the opposite of INBREEDING.

Outcrossing similar to OUTBREEDING, but implying a deliberate mating between different strains, breeds or lines of a species.

Ovary organ which produces the OVUM. In vertebrates it also produces sex hormones.

Overhead cost in animal production, the costs of building, heating, labor etc, as contrasted with the costs of food for animals or purchasing the animals.

Oviduct tube carrying the OVUM from the OVARY.

Ovulation the release of mature eggs from the ovary of a female animal.

Ovum (plural ova) unfertilized egg-cell.

Pastoral of land used for pasture.

Pasture the growing herbage eaten by cattle. Also a piece of land covered with this. Also (verb) to feed animals on this.

Pedigree genetic origin, line of succession.

Pelagic the upper part of the open sea, above the BENTHIC zone.

Pelt the skin of a mammal with short wool or hair on it.

Pendent or **pendulous ear** ear that hangs down.

Phylum a taxonomic division comprising a number of CLASSES.

Pied a coat or PLUMAGE pattern displaying two or more colors, one white and another often black. May be in an irregular or regular pattern.

Pigment a chemical compound which gives color to skin, eggs, feathers or hair.

Plankton mostly very small animals and plants of sea or lake which float or drift almost passively. Of great ecological and economic importance, providing food for fish and whales.

Plumage the feathers of birds.

Polled with horns removed or not developed.

Pollination transfer of pollen from anther to stigma in a flower or flowers in order to fertilize.

Polygyny a mating system in which a male mates with several females during one breeding season (as opposed to polyandry, where one female mates with several males).

Population a more or less separate breeding group of animals. Also, the total number of individuals counted in a given area.

Pork the flesh of a pig other than that used as BACON.

Poult the young of the domestic chicken, duck, goose, turkey, pheasant, guinea-fowl, or various game birds.

Poultry collective name for domestic chickens, ducks, geese, turkeys, quail and closely related species.

Pre-adaptation an ADAPTATION which predisposes a species to occupy a new NICHE successfully even though it had not previously undergone SELECTION for this new role.

Predator an animal which forages for live prey.

Primitive trait a trait present at a very distant time in the evolutionary history of a related group of animals.

Progeny offspring and future descendants.

Prolificacy measure of number of offspring produced.

Proteins large organic molecules made up of AMINO ACIDS. Present as structural components of cells and tissues and as ENZYMES. Along with FATS and CARBOHYDRATES they are one of the three classes of energy-providing nutrients.

Puberty the attainment of sexual maturity. This involves maturation of the primary sex organs (OVARIES and testes) and of secondary sexual characteristics, such as song in male birds, and raised leg urination by male dogs.

Purebred bred from genetically similar ancestors. Not the result of crosses between breeds.

Quarter The leg of a quadruped together with the adjoining parts of the body used for locomotion. The hindquarters are the two rear quarters.

Rabies a viral disease that can affect most species of mammal including man and is almost always fatal. Symptoms vary between species, but death is ultimately caused by damage to the nervous system. In most countries, pets and livestock are potentially at risk of infection from wild species such as the fox and racoon, and compulsory vaccination laws are widespread. In countries such as Australia and the UK, which are rabies free, quarantine of imported dogs and cats is compulsory to prevent introduction of rabies.

Race a taxonomic subdivision of a species whose members are distinguishable from the rest of that species.

Receptive of a female, willing to accept sexual advances of a male.

Recessive of a GENE or trait: one whose characteristics are only displayed if it is inherited from both parents and therefore not masked by a DOMINANT GENE.

Register to record the particulars of an animal's ancestry and birth with a BREED SOCIETY or other body.

Rickets a disease, due to dietary deficiency of vitamin D, characterized by softening of the bones, especially of the spine, and consequent distortion, bow-legs and emaciation.

Roan a type of coat found in some horses and cattle consisting of a basic coat color thickly interspersed with a second color, eg blue roan is black mixed with white, strawberry roan is red mixed with white and gray.

Roughage indigestible plant fiber in food which simulates passage of food through the intestine.

Ruminant a mammal with a specialized digestive system typified by the behaviour of chewing the cud. The stomach is modified so that vegetation is stored, regurgitated for further chewing, then broken down by symbiotic bacteria. This process is an adaptation to digesting the cellulose of plant cell walls.

Rumination the action of chewing the cud.

Rump the tail end, posterior or buttocks of a mammal. In birds, it is the area just above the base of the tail. A goose rump in a dog or a horse is one that slopes very sharply from the CROUP.

Sanga groups of cattle breeds which originated from crosses between humpless longhorns and ZEBU cattle. They have long horns and a small hump in front of the withers.

Saturated fat FAT without double bonds in its molecular structure.

Scapula the shoulder blade.

Scent gland an organ secreting odorous material with communicative properties for SCENT MARKS.

Scent mark a site where the secretions of SCENT GLANDS or urine or feces are deposited and which has communicative significance. Often left regularly at traditional sites which are also visually conspicuous. Also, the "chemical message" left by this means: and (verb) to leave such a deposit.

Scrub a vegetation dominated by shrubs—small woody plants usually with more than one stem. Naturally occurs most often in ARID forest or grassland but often artificially created by man as a result of forest destruction.

Seasonal breeder animal that breeds at only one period or season in the year, as opposed to a continuous or cyclic breeder.

Sedentary occupying a relatively small home range, and exhibiting weak dispersal or migratory tendencies.

Seed oysters oysters used for egg production.

Seed prawns prawns used for egg production.

Selection the process by which certain individuals and their unique characteristics are favored over others through more successful breeding and better survival of offspring. These characteristics thus come to be represented in a greater proportion of the POPULATION. See also ARTIFICIAL SELECTION. NATURAL SELECTION.

Sericulture the production of raw silk and the rearing of silkworms for this purpose.

Sex cycle a recurrent period of alternating sexual and nonsexual activity. The ESTRUS cycle is an example.

Sex-linked a form of inheritance where a characteristic is solely or more commonly found in only one sex. This is because the gene for this trait is carried by only one of the sex chromosomes (see X CHROMOSOME: Y CHROMOSOME).

Sheath a tubular fold, including that into which the penis is retracted.

Silage FODDER preserved in a silo or pit, without previous drying.

Sire the male parent of an offspring. Commonly used as a term for a breeding STALLION.

Solitary living on its own, as opposed to a social or group-living lifestyle.

Sour cream partially fermented cream.

Sow a full-grown female pig whose first litter has been weaned.

Species basic division of biological classification, subordinate to GENUS and superior to subspecies. In general a species is a group of animals similar in structure and which are able to breed and produce viable offspring.

Sperm oil an oil found in various species of whales, including the Sperm whale.

Spina bifida congenital defect in which the backbone fails to fuse and surround the spinal cord. Seen in some Manx cats.

Spur the sharp projection on the leg of some game birds: often more developed in males and used in fighting. Also found on the carpal wing joint of some other birds.

Stallion an uncastrated male horse.

Steer a castrated male BOVINE.

Steppes level, grassy plains largely devoid of trees as found in part of SE Europe and Asia.

Sterile unable to reproduce sexually. Also free from living micro-organisms.

Stock the animals on a farm: collective term for horses, cattle, sheep etc bred for use or profit. Fat stock are kept for meat: breeding stock are kept for breeding.

Stone Age period of human culture when tools were made of stone or bone. In Europe, this was succeeded by the Bronze Age about 3000-2000 BC.

Strain a group of related individuals who resemble each other in both appearance and genetic constitution. A subdivision of either a SPECIES or a BREED, but not recognized for any show purpose.

Stud a male animal specially kept for breeding. Also, a place where STALLIONS are kept for breeding. A stud fee is the money paid by the owner of a female to the owner of a male used for breeding.

Stud book record containing details of the PEDIGREE of horses.

Sty an enclosed space where pigs are kept, usually a low shed with an uncovered forecourt.

Subcutaneous fat the layer of fatty tissue immediately underlying the skin.

Sub-orbital situated below or under the orbit of the eye.

Subordinate individual which has a lower status than other members of its social group. (See DOMINANT ANIMAL.)

Subspecies a recognizable subpopulation of a single species, typically with a distinct geographical distribution.

Swine pig(s).

Taiga northernmost coniferous forest with open boggy areas.

Tallow the fat or adipose tissue of an animal, especially that obtained from the parts about the kidneys of ruminating animals.

Tame gentle and tractable to humans.

Tankage the residue from the tanks in which fat etc has been rendered, used as a coarse food and as manure.

Tanning changing skin into leather by chemically altering its PROTEINS, especially by soaking in tannin.

Taxonomy the science of classifying organisms. It is very convenient to group together animals or plants which share common features and are thought to have common descent. Each individual is thus a member or a series of ever broader categories (individual – species – genus – family – order – class – phylum) and each of these can be further divided where it is convenient (eg subspecies, superfamily).

Territorial behavior behavior used by an animal to delineate or defend an areas from intruders either on behalf of its or for a social group. Includes birdsong, SCENT MARKING and visual display.

Thorax chest, part of trunk between neck and abdomen.

Tom the male of various mammals and birds, eg cat, turkey.

Trace element a chemical element which must be available to an organism for its normal health although it is necessary only in minute amounts.

Transhumance the seasonal moving of livestock to regions of different climate.

Tribe a term sometimes used to group certain SPECIES and/or GENERA within a family.

Tuberculosis a bacterial disease affecting birds and mammals which is marked by the pressure of small round swellings or tubercles in internal tissues, especially the lungs.

Tundra barren treeless lands of the far north of Eurasia, North America and Arctic islands. Dominated by low shrubs, herbaceous perennials, with mosses and lichens.

Type the characteristics which distinguish a breed as detailed in its standard written by the governing body for the breed, similar to CONFORMATION. Also a variety of animals sharing similar characteristics or uses but not recognized as a BREED. For cagebirds an equivalent term to "breed," but restricted to differences other than color.

Umbilical fold a fold of skin over the umbilicus.

Upright shoulder an insufficient angulation of an animal's shoulder blades compared with the breed standard.

Urea main excreted product of PROTEIN break-down in mammals.

Uterus in mammals, the womb of the female where the fetus develops and is attached to the mother by the placenta. In birds an alternative name for the shell gland.

Utility breed a BREED which is of commercial importance rather than being just ornamental.

Variability the showing of a considerable range of appearance for a particular character.

Variation differences in character of organisms in a species or group.

Variety a group of individuals which resemble each other in appearance, often color. A subdivision of BREED.

Vector an individual or SPECIES which is an intermediary host or agent in the transmission of a disease or parasite from one organism to another.

Vegetarian one who lives wholly or principally upon vegetable foods.

Velvet antler the soft downy skin which covers a deer's antler while in the growing stage.

Venison the meat of a deer.

Vertebrate an animal with a backbone: a division of the PHYLUM Chordata.

Viscera the soft contents of the principal cavities of the body, eg lungs, stomach, intestines.

Vitamin any of several organic substances, distinguished as vitamins A, B, etc, occurring naturally in minute quantities in many foodstuffs and regarded as essential to normal growth, especially through their activity in conjunction with ENZYMES in the regulation of METABOLISM.

Vocalization sounds or calls, produced in mammals by the vocal chords of the larynx, produced in birds by an organ called the syrinx which branches from the trachea.

Wattle a fleshy lobe (usually brightly colored) pendent from the head or neck of certain birds, eg chicken, turkey.

Well-sprung having a rounded, well-formed rib cage.

Wild not domesticated: intractable, opposite of TAME. Also, a habitat that has had little or no disturbance from the activities of man.

Wild type term applied to a GENE or an individual which corresponds to that found most frequently in the WILD.

Withers the highest part of a quadruped's back, lying between the shoulder blades, just behind the neck. "Mutton withers" are broad and rounded as in sheep.

Wool soft undercoat of hairy animals, especially the artificially selected fleece of wool breeds of sheep.

X chromosome the longer of the two sex chromosomes. In most mammals females have two X chromosomes, and males an X and a Y CHROMOSOME. In birds the males may be XX and the females XY depending on species.

Y chromosome the smaller of the two sex chromosomes, therefore including fewer GENES than the X CHROMOSOME with which it is paired (see SEX-LINKED).

Yogurt a sour fermented liquor made from milk.

Zebu humped cattle first domesticated in India.

INDEX

Picture Acknowledgments

Key t top. b bottom. c center. l left. r right.

Abbreviations A Ardea. AN Agence Nature.
ANT Australasian Nature Transparencies. BCL
Bruce Coleman Ltd. CHL Camerapix Hutchison
Library. NHPA Natural History Photographic
Agency. OSF Oxford Scientific Films. PEP Planet
Earth Pictures/Seaphot. RHPL Robert Harding
Picture Library. SAL Survival Anglia Ltd.

Cover A/Wilhelm Moller. 1, 2–3 Tony Stone
Associates. 4–5 BCL/Mark N. Boulton. 6–7
BCL/Nicholas Devore. 8–9 RHPL. 10–11
CHL. 12–13 ICCE/M. Boulton. 15 RHPL. 16b
OSF/P.K. Sharpe. 16t CHL. 18 R.
Fletcher/Swift Picture Library. 19 T. Owen
Edmunds. 21 BCL/M. Gunther. 22
D. Hosking. 23 D. Broom. 24 A/P. Steyn.
24–25 A. Bannister. 26–27 BCL/

C. Davidson. 29 BCL/Bauer. 30 NHPA/E. Janes.
33 Aquila/Lane. 34–35, 35t RHPL. 36
T. Owen Edmunds. 36–37 S. Hall. 38t A/
P. Morris. 38b BCL/M. Freeman. 39 NHPA/
S. Dalton. 40–41 BCL/J. Wright. 46 CHL/
S. Porlock. 47t SAL/M. Kavanagh. 47b,
48–49, 57 CHL. 58–59 C.A. Henley. 60
B. Hawkes. 61 CHL. 62–63 BCL/
C. Bonnington. 63t BCL/H. Jungius. 64–65
RHPL. 66 BCL/E. Dragesco. 67 A/
J.P. Ferrero. 69t A/F. Gohier. 69b ANT/R.
& D. Keller. 74t A/Bomfords. 74–75 NHPA/
A. Bannister. 80 G. Bateman. 81 A/R. Porter.
82–83, 82b BCL. 83 M. Holford. 86b
A. Mowlem. 86–87, 87b CHL. 92–93 A/
J.P. Ferrero. 93t CHL. 97 Naturfotograferna/
T. Hagman. 98 R. Luxmoore. 99 CHL. 100
W.N. Bonner. 101 A/Y. Arthus-Bertrand.

103 BCL/M. Boulton. 104t BCL/C. Marigo.
104b A. Beaumont. 106–107 FL. 109,
112–113 A/J.P. Ferrero. 114t BCL/
C. Marigo. 114–115 Aquila. 117 BCL/
B. Hamilton. 118t BCL/L.L. Rue. 118b A/
J. Bailey. 119 RHPL. 120–121 BCL/
N. Tomalin. 124 AN/J.P. Ferrero. 126
Biofotos/H. Angel. 127 RHPL. 129 BCL/
W. Ferchland. 130 BCL/D. Goulston. 131
BCL/K. Gunnar. 132 RHPL. 134–135
Biofotos/H. Angel 135b PEP/Lythgoe. 138
BCL/G. Williamson. 141 OSF/G. Bernard.
142 J.B. Free. 143 A. 145t BCL/Prato. 145b
AN/Lanceau. 146–147, 147b Aspect/
P. Carmichael. 148 BCL/M. Viard. 149t
BCL/M. Freeman. 149b A.C. McDougal. 150
ANT/P. Krauss. 151 A/R. Waller.

Artwork

All artwork ©Priscilla Barrett unless stated
otherwise.
Abbreviations SD Simon Driver. NM Nick
Mynheer.

11, 14b NM. 14t Wayne Ford. 17, 18, 21,
27, 31, 34 NM. 37 SD. 56, 57, 61, 95t,
105, 110 NM. 111, 112, 116, 117 Equinox.
125 NM. 127 SD. 128, 129 Mick Loates.
131 SD. 132, 133, 136–137 Mick Loates.
138, 139b Rob van Assen. 139t NM. 140
Richard Lewington. 143 SD. 144 NM. Maps
and scale drawings SD.